SSADM VERSION 4: A User's Guide

THE McGRAW-HILL INTERNATIONAL SERIES IN SOFTWARE ENGINEERING

Consulting Editor

Professor D. Ince
The Open University

Other titles in this series

Portable Modula-2 Programming	Woodman, Griffiths, Souter and Davies
SSADM: A Practical Approach	Ashworth and Goodland
Software Engineering: Analysis and Design	Easteal and Davies
Introduction To Compiling Techniques: A First Course Using ANSI C, LEX and YACC	Bennett
An Introduction to Program Design	Sargent
Object Oriented Databases: Applications in Software Engineering	Brown
Object Oriented Software Engineering with C + +	Ince
Expert Database Systems: A Gentle Introduction	Beynon-Davies
Practical Formal Methods with VDM	Andrews and Ince
Introduction to VDM	Woodman and Heal
A Structured Approach to Systems Development	Heap, Stanway and Windsor
Software Engineering Environments	Brown, Earl and McDermid
Introduction to Software Project Management	Ince, Sharp and Woodman
Systems Construction and Analysis: A mathematical and logical framework	Fenton and Hill

SSADM Version 4: A User's Guide

Malcolm Eva

McGRAW-HILL BOOK COMPANY

London · New York · St Louis · San Francisco · Auckland · Bogotá
Caracas · Hamburg · Lisbon · Madrid · Mexico · Milan · Montreal
New Delhi · Panama · Paris · San Juan · São Paulo · Singapore
Sydney · Tokyo · Toronto.

Published by
McGRAW-HILL Book Company Europe
Shoppenhangers Road · Maidenhead · Berkshire · SL6 2QL England
Tel 0628 23432   Fax 0628 770224

British Library Cataloguing in Publication Data

Eva, Malcolm
  SSADM version 4: a user's guide.
  I. Title
  658.40380285421
  ISBN 0-07-707409-2

Library of Congress Cataloging-in-Publication Data

Eva, Malcolm
  SSADM version 4: a user's guide / Malcolm Eva.
    p.   cm.
  Includes bibliographical references and index.
  ISBN 0-07-707409-2
  1. System.   2. System analysis.   I. Title.
  QA76.9.S88E9   1991
  004.2'1–dc20   91–26225 CIP

2345 CL 95432

Typeset by Computape (Pickering) Ltd, North Yorkshire
and printed and bound in Great Britain by Clays Ltd, St Ives plc

Contents

Preface

Scope

Since this book's conception, its scope has changed and narrowed to the form it appears in today. Originally, it was to be a detailed exposition on Version 4 of SSADM, descriptions of how it could and should be implemented in organizations, and finally a detailed discussion of its strengths and weaknesses, comparing it with other systems analysis methods.

When Version 4 was launched, in June 1990, it was clear that the changes were so extensive that the book could not keep its intended scope and stay a commercial proposition. So, I have cut ambitious plans to produce *the* standard volume of methodology and confined myself instead to a description of SSADM Version 4, pure and simple. I did this reluctantly at first, but by the time I was halfway through the draft, I felt only relief that I had not let myself in for the rest as well.

My first feelings about Version 4 were ambivalent: I had not expected it to change as much from Version 3 as it has, and was furious that much cunningly preprepared work for this book, based on Version 3, all had to be scrapped. There was a lot that was new; that was daunting. Some things that I knew well had changed, or lost importance; that made me a little indignant on their behalf. New complexity seemed to have been built in for no good reason, and I thought that it would be a far harder method to sell to organizations.

However, further examination of the material changed by views—not in one bound, but gradually and surely. The weaknesses in Version 3 had been addressed: Dialogue Design and Physical Design were two such areas. Both have been changed completely, and given a much better chance of achieving their objectives. The analyst is not locked into views of the current system, as was the case with Version 3; rather, the emphasis is on identifying and meeting the requirements for the new system.

Any project undertaken using SSADM could not now put a half-trained team of analysts on to a job, with virtually no Project Management support and expect it to deliver the goods, as I have seen happen with several methods, including SSADM Version 3. When such projects failed, SSADM was usually blamed. This was unfair, always, because the team were not allowed to use SSADM techniques properly, were not given reasonable estimating guidelines, and received no clear leadership. Version 4 has met that particular dragon and disposed of it with its new structure and mandatory interfaces with Project Management.

The deeper I looked into the new face of SSADM the more I saw the strength of the new approach. The increased rigour in the definition of all SSADM products and cross-checks has led to a more reliable end-product, while at the same time the modular structure and the stress on non-SSADM activities has led to greater flexibility, both in the scope of projects undertaken and in the emphasis each separate installation might place on particular aspects, such as ergonomics, or the human–computer interface.

The packaging of the enhanced method embraces a comprehensive manual, a SSADM dictionary and two entity–relationship–attribute meta-models of the method. This book does not attempt to outdo that packaging; its purpose is to condense and clarify the new structure of SSADM, and, using a continuous worked example, demonstrate the techniques employed.

I have not gone into the history of SSADM in any detail; neither have I rehearsed the justification for using structured methods of analysis and design. I have started from the assumption that a reader understands already that a method for systems analysis and design is preferable to the traditional seat-of-the-pants approach. What I expect a reader to ask is not 'Why should I use a structured method?' but 'What is SSADM Version 4?' That is the question I hope this book will satisfy.

Structure of the book

The book is in three parts: Part 1 introduces SSADM, and describes the new Structural Model. Each of the five modules that make up the core method has a chapter of its own that describes the inputs, outputs, stages, steps and tasks employed.

Part 2 of the book describes the techniques. Each technique has a chapter to itself. A possible exception occurs in the description of Entity/Event Modelling, Chapter 13. Two techniques are described here: Entity Life History Analysis and Effect Correspondence Diagrams. Each is a facet of Entity/Event Analysis, but the techniques themselves are very different. Part 2 begins at Chapter 8, with a preview of the different techniques, and how they fit together.

Part 3 comprises a chapter on applying SSADM within an organization, and then a glossary of SSADM terms. The glossary is more substantial than that in many books, which tend to list a dozen common words and feel their duty is done. The source for this glossary is largely—but not exclusively—the dictionary volume in the *SSADM Reference Manual*. As with any specialism, the field of SSADM is rich with jargon, some of which is familiar from other contexts, and therefore possibly misleading. I have tried to incorporate all terms that are likely to be used in the course of a SSADM project. I am sure that many have slipped through the net, even so.

The exercises

There is a difficulty in choosing which techniques to reinforce with exercises in a book such as this. Some 'techniques' in SSADM are not so much techniques as continuing activities. Dialogue identification and design is a suite of activities like this. Requirements Definition is another.

Some techniques, such as Function Definition, have such large inputs to carry out, that to include them may require the reader to examine more detail than is practical in a book-based exercise.

Accordingly, the exercises I have set may seem to stay in already well-trodden ground. I have selected the three views of data as the basis for the practical work, because this is still at the heart of SSADM, and the skills needed to practise the method are still oriented towards these. The new aspects of these views, Enquiry Access Paths and Effect Correspondence Diagrams, are included. Relational Data Analysis, although it is less prominent in Version 4, is nevertheless extended to include Boyce–Codd, Fourth and Fifth Normal Forms. Practice at these is therefore offered.

The omissions are Function Definition, as explained above, Logical Database Process Design (which takes the I/O Structures from Function Definition as its input), Dialogue Design (the same) and Physical Design. Physical Design is excluded because as with previous versions it is only a generic guide. If Physical Design, either data or process, were performed in precisely the same way regardless of implementation environment, an exercise would be fruitful. As it is, I see little use in exercising a non-existent DBMS in preference to the real thing.

Acknowledgements

It is impossible to write a book of this nature without the help and support of others, even though it is for the most part a solo—and gruelling—effort. I would like to express my thanks to the following people who, in one way or another, helped me to reach this stage: Andy Ware of McGraw-Hill Book Company, for his constant encouragement when I feared I was losing my way; Caroline Ashworth of AIMS Systems and Duncan Quibell of Brighton Polytechnic for giving the manuscript a much-needed technical review, and making valuable suggestions as to accuracy and presentation; colleagues at Telecom ACT for general encouragement and specific advice, particularly Malcolm Pither, John Rogers and Rob Bailey; Diane and Natalie for patiently putting up with my vanishing from normal homelife every night for several months while I wrote this book, and supporting me all the while; Natalie for her help with the typing.

I am grateful to Learmonth and Burchett Management Systems (LBMS) for their permission to use and describe the Jackson structure notation, which is their copyright, and which is central to several graphical techniques of SSADM Version 4.

PART ONE   Structure of SSADM

1. Introduction to SSADM

1.1 Aims of chapter

In this chapter I shall discuss the following topics:
- the history of SSADM,
- the principles of SSADM and
- the drive towards Version 4 of SSADM, launched in 1990.

As the thrust of this book is towards a description of SSADM in its current state, the reader will find that almost all subsequent chapters deal with the structure and techniques of SSADM Version 4. I have, as stated in the preface, narrowed the scope of the text to description, rather than evaluation or a discussion on comparative methods.

Although this book may be used as a pocket manual by practitioners, I think it is important for the reader not only to see SSADM in a context, but also to understand both why it was developed and the philosophy and principles behind its shape, as well as appreciating how, as a method, it has been enhanced to keep pace with the changes in the world.

1.2 Background of SSADM

History

Structured Systems Analysis and Design Method (SSADM) has been used by the Government in computing since its launch in 1981. It was commissioned by the Central Computing and Telecommunications Agency (CCTA) in a bid to standardize the many and varied IT projects being developed across Government departments. The CCTA investigated a number of approaches, before accepting a tender from LBMS to develop a method.

Since its first appearance and adoption, SSADM has undergone changes, partly to remedy perceived weaknesses in the method, but mostly as a response to developing technology and changes in the business world. In 1986 Version 3 was launched, its main change being the introduction of a Dialogue Design technique to address the growth in online processing, a development that earlier versions had not tackled.

Open standard

Although SSADM was designed for the Government, the CCTA were keen for it to become an open standard, freely available to other interests in industry, or in the academic community, for instance. A benefit from this is that industry would then have an interest in producing support for the method. This support would principally be in terms of training and automated tools. This, in fact, has been mutually beneficial to the method and to industry, and SSADM is now used widely throughout industry, as well as by the Government.

SSADM now

The current state of SSADM is Version 4, released in 1990 after three years of work in drawing up a statement of requirements, tendering and finally developing the new method.

1.3   SSADM philosophy

Throughout SSADM's history, certain features have been central to it. These have been carried forward through all versions, and are still at the heart of Version 4. New emphases have been added, but they supplement rather than supersede the central features. These constant features can be summed up under three headings:

- three views,
- avoid lock-in,
- user involvement.

These headings have been well rehearsed since the method's first public appearance in 1981, so to maintain continuity, I shall define briefly what each means. We see later, that each has, in fact, since been enhanced by a combination of greater rigour and pragmatism.

Three views

The views in question are views of the System Data. They represent function, structure, and the effects of time, as shown in Figure 1.1. Each of the various philosophies of systems analysis emphasizes a particular approach to the system: structured analysis, as propounded by Yourdon, De Marco and others, explores the functionality of a system, and techniques such as data flow diagrams investigate the transformation of system data as a result of inputs from outside the system boundary.

This can be seen as a form of information plumbing, with pipelines (or data flows) carrying the data from sources to sinks, via stations that carry out some sort of transformation on that data.

Another view, the data analysis view, says it is all well and good looking at the functionality of systems, but the important thing is the underlying structure of the data. To understand that, we have to build an entity model (or, in SSADM terms, a *Logical Data Structure*). Once the data structure is in place it will stay constant, regardless of what else happens in the business. Procedures and functions may change, after all, but the data must always be there to support the business activities, whatever they may be.

The third view says that yes, it is fine recognizing data structures, and seeing how the plumbing carries data from one part of the system to another, as and when it is needed, but both models are static, giving snapshots of possibilities. What is really

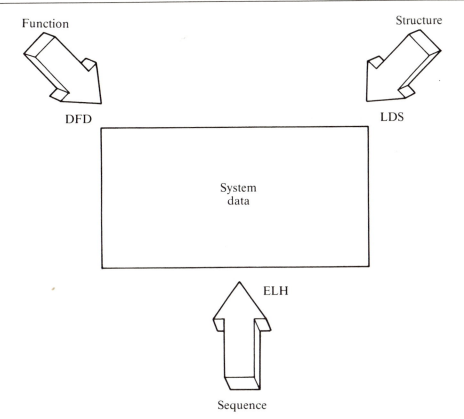

Figure 1.1

needed is a way to model business and integrity rules as they affect the data. Accordingly, we must model the effects of time on our data, and show how each item can be brought into being, how it can be amended and how it can be removed from the system, according to the business rules of the application area. This third view gives us the *Entity Life History* (ELH).

Each of the philosophies places one of these views above the others in importance. SSADM instead gives each an equal importance, and makes them complement each other. A series of cross-validation checks between them ensures that at the end of analysis, as rich a picture as possible is gained of the system.

The Data Flow Diagrams (DFDs) act as a check on the *Logical Data Structure* (LDS), to show that each data item is created and amended at some time. The LDS ensures that the data is present in the system, to allow the function processing shown on the DFD. The ELH subsequently identifies all of the events that impact on the system, causing changes to the state of the data. At this point, omissions from earlier investigations are frequently uncovered, and the picture expanded. Integrity rules for making amendments to data are discovered and built in.

The three views of SSADM ensure that a detailed analysis of the target systems is carried out.

Avoid lock-in

It could be argued that the very early forms of systems analysis were practised by salesmen from the large computer manufacturers. Clients with an information systems requirement would find their needs being adapted to suit the prescribed solution, rather than the reverse.

The problems arising from this are obvious: the technology is driving the solution, not vice versa. Should the system need to be enhanced or modified later, the client is locked into the original supplier, and must have the extended solution dictated by the constraints of the technology once again. Should the supplier go out of business, or the hardware/software become obsolete, the client really is in trouble.

SSADM approaches the problem from the other end. The nature of the problem is analysed, and a specification for a solution is drawn up, independently of any one implementation environment. A Logical Design is produced for both data and processing, which can be implemented on a selected environment or ported between environments with no adjustment or alteration. The Physical Design would need amending, since that is implementation-specific, but the Logical Design, which meets the clients' needs, is constant. Enhancements are carried out on the Logical Design, which is flexible enough not to need a re-specification in the process.

User involvement

Aesop tells of a fox who was attracted by grapes growing high up a vine. In fact, the grapes were so high that however high the fox jumped, they were always just out of reach. About to give up, and declare that they were too sour anyway, the cunning fox saw a systems analyst walking down the road. As this was a tale told in Ancient Greece, the systems analyst had not been trained in SSADM, and regarded himself as an expert on all matters on which he was consulted.

The fox explained his problem: he could not reach the grapes, which he desired to eat. 'Right you are', said the systems analyst, 'Leave it with me, and by nightfall the grapes will be in your reach. I shall have two of them as my fee.'

The fox, being cunning, asked the systems analyst if any more information were needed, or could he help, but the systems analyst said, 'I can cope now, thank you very much, run along till nightfall'. When the fox returned at nightfall, the systems analyst said that all was nearly ready, and could he pop back in a couple of hours. After two more returns, at dawn the systems analyst announced that the project was complete, and the fox could go-live and collect his grapes now.

The fox looked, and saw that the systems analyst had built a scaffolding around the vines, for the fox to climb up. 'I thought you were going to shin up yourself and pick them for me', said the fox, aggrieved. 'I can't climb up a scaffolding, foxes aren't built for that. Anyway, there isn't even a ladder to get me to the first platform. I'm not accepting this system.'

The analyst pleaded in vain that he had carried out his terms of reference, and that the fox had not asked him to pick the grapes himself.

The fox was so upset that he even contemplated eating the systems analyst, but thought the analyst would probably be too sour, so he slunk off, vowing never to ask an expert for help again.

Had the systems analyst been SSADM-trained, he would have understood that the user is the expert in the applications area. He would have ensured that the fox's requirements were thoroughly analysed and understood before he designed a solution, and would have offered the fox a set of possible solutions so that the fox himself could have chosen the way to reach the grapes. At every point in the project where a part of the investigation or specification was completed, the fox would have been asked to inspect the results and approve them, before the next stage was begun. At the end, the fox would have had his requirements met in full, and the systems analyst would have been paid.

Where this fable departs from reality is that, too frequently, the systems analyst would have been paid before the user found that the system was not what was wanted.

1.4   The drive to Version 4

There is a well-known saying in engineering circles: if it works, don't fix it. In that case, if SSADM is used widely, and is becoming a de facto standard, why produce such a radical change as Version 4, at considerable expense?

The answer lies in two areas: user experience, and the developing technological environment:

User experience

This falls into another two broad areas, experience of SSADM, and experience of changes in the business IS world.

The experience of SSADM, while generally a happy one, has prompted some criticisms of aspects of the method. Weaknesses included such areas as the translation of a Logical Design to a Physical Design. The guidelines were very thin, and SSADM seemed to peter out after Stages 4 and 5 were complete.

The online processing as handled in Dialogue Design was not felt to address the issue in sufficient depth. What the technique achieved was to specify the User Interface, in logical terms, but it was not a part of design, and had little to do with dialogues. A more rigorous approach to the subject was requested, and so became a part of the requirements for Version 4.

Another, more general comment, was that SSADM spent too much time looking at the current system, with its attendant problems and the illogicalities that had evolved. The thrust needed to be towards the requirements for the new system.

There were other comments about how the method was used, and more importantly, how it could be misused, that were fed into the statement of requirements.

Users were subject to other influences, however. The criticisms of SSADM were not damning ones, and only required a patch or two. The influences that caused the radical re-examination were to do with changing experiences in development of Information Systems. Users were being given opportunities to see their system grow with prototypes; users now had greater choice when buying in services from third parties, and were able to make that buying-in decision at more than one place in a development project.

There has been a greater interest in end-user computing, whereby business users are able to build their own applications apart from the IT function.

SSADM did not cater specifically for these and other new expectations from the user. There was a requirement for some form of guidance as to how these could be utilized under a SSADM umbrella.

Technological developments

One of the reasons that the old tea-clippers and wooden warships gave way to steamships, ironclads or even aeroplanes was quite simply that these things came into existence, and did the job of transport/sea warfare better than the old equipment. Computing technology is developing faster than the old marine transport technologies did; whereas earlier versions of SSADM suited the then current state of technology, there have been many advances since.

On the systems development front, there has been a growth in the production of 4GLs and application generators (AGs). While the products of SSADM Stages 4 and 5 could be fed into an AG, it was never explicitly stated how and where. A tighter definition of the interfaces was requested.

Distributed systems is another area that has developed rapidly. Orthodox structured methods did not appear to cater for this form of architecture. In fact, SSADM could be tailored intelligently to cope with distribution, but again, this was never described explicitly.

A new technology that is moving from the academic arena into commercial development is object-orientation, which as a philosophy is a departure from orthodox database/data-processing practice. There was a call for SSADM to address this new approach, as well.

Technology continues to move forward, and business environments continue to evolve. The two sets of dynamics contribute to a 'push-me, pull-you' effect on a method, and insist that it is upgraded to keep pace with the new demands, however satisfactory it was at the earlier stages of its life.

Design criteria

The Design Authority Board, the body responsible for maintaining SSADM, established certain base criteria for the development consortium. The principal aims were to:
1. Enhance the ease of use for practitioners.
2. Encourage computer assistance, and help move SSADM away from being a paper-based method.
3. Increase the rigour of the techniques.
4. Improve the quality of both SSADM and SSADM products.
5. Maintain SSADM's flexibility.

The CCTA also made some stipulations. They were, primarily, to:

1. Minimize the changes for the SSADM user.
2. Maintain the three views of system data, as described earlier.
3. Maintain the separation of Logical and Physical Design.

There were other design criteria laid down by the Design Authority Board, but the ones listed above were the essential governing constraints.

The new SSADM

The result of these inputs was a release of SSADM that met nearly all of the issues it was asked to address. What emerged was a method that:

- had a modified structure (see Chapter 2);
- introduced explicit interfaces with ancillary activities to support the analysis and design activities (see Chapters 2 and 21);
- placed a new emphasis on requirements, and less on the current system (see Chapter 15);
- placed a new emphasis on functions from the business user's perspective (see Chapter 11);
- revised old techniques, making them more rigorous, and introduced a number of brand-new techniques (see Chapters 8 to 20);
- replaced the activity-driven view of the method with a product-driven view (see Section 1.5); and
- introduced mechanisms for making the planning and control of a project easier and more reliable (see Chapter 21).

The rest of this book describes the philosophy, structure and techniques of SSADM Version 4.

1.5   A product-based approach

The new approach to SSADM, while radically different at first sight, shows itself, on further inspection, to be close to the original. What it has done is to make explicit what was implicit before, and because it was implicit, was often overlooked or omitted. The need for a clear interface for Project Management has always been there, but now the 'hooks' are explicitly defined, as are the interfaces for such activities as Risk Assessment, Configuration Management, Capacity Planning, Quality Control, etc.

No IS development project should ignore these activities, but unless they are built into a method, the temptation to take a short cut for the sake of showing a system in operation seems irresistible.

Apart from an altered structure, the approach that SSADM has taken to build these in is to change its orientation from activity based to product based.

What, when, how

The practitioner is now tasked with producing predefined deliverables. The end of an activity is reached when this deliverable is accepted by the project manager (or module manager, or stage manager). The new structure of SSADM tells the project team *what* is to be produced, *when* it is to be produced, and *how* it is to be produced.

WHAT
SSADM now includes a dictionary, to define every product that comes from an

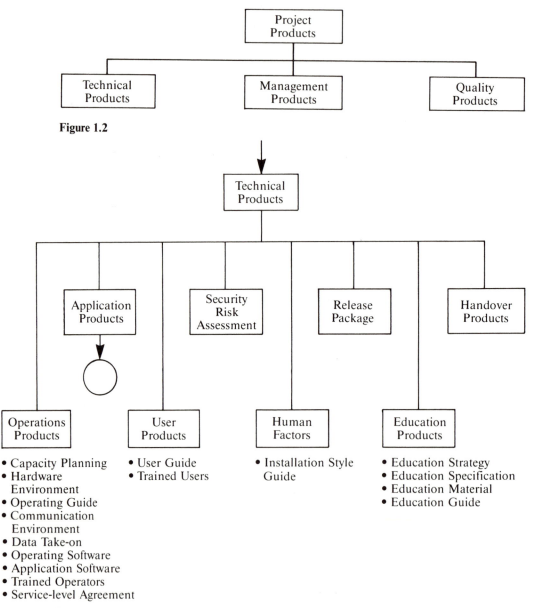

Figure 1.2

- Capacity Planning
- Hardware Environment
- Operating Guide
- Communication Environment
- Data Take-on
- Operating Software
- Application Software
- Trained Operators
- Service-level Agreement

- User Guide
- Trained Users

- Installation Style Guide

- Education Strategy
- Education Specification
- Education Material
- Education Guide

Figure 1.3

activity. There is a hierarchy of such products, which fall under three broad headings: *technical, management* and *quality*. Figures 1.2–1.6 show the hierarchy of products under those headings.

The lists are not definitive: local standards or practices may require additional, or alternative, deliverables. What is important, is that every product which is to be delivered to the Project Management is to be specified under its appropriate place on the tree.

The products from the *Applications Product* node are produced by the SSADM practitioners. I have not enumerated the products for each Module of the applications node: they are described in Chapters 3–7, under the *output* headings of the Module descriptions.

Management will require the products from management node. Again, not all of the leaves have been specified in full: management will need to identify and define each separate product they expect to see from each leaf.

The definition of products is carried out before the event, not after, or at the same time. The description of each required product is a key input to each activity (see Chapter 2).

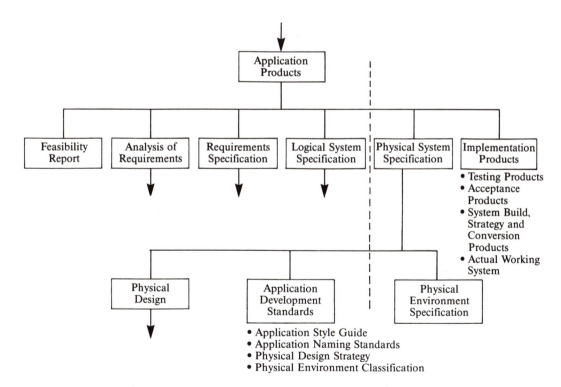

Figure 1.4

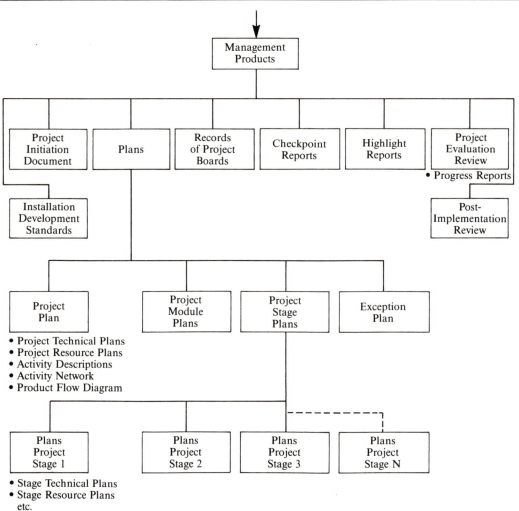

Figure 1.5

Product specification

Each product on the hierarchy is defined in the following terms:

Title   A simple identifier, which is accessible to non-technical people as well as the technical specialists.

Purpose   A brief statement of what the purpose of the product is, e.g. to communicate the analyst's understanding to the User, or to show the layout of a screen used in a dialogue.

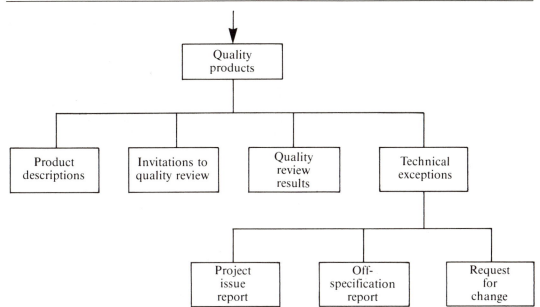

Figure 1.6

Composition   A list and description of all the component parts of the product. If it is a report, or form based, a design for the form should be included, showing project header information, version numbers and release dates for Configuration Management purposes, as well as space for every item of information required. The *SSADM Reference Manual* provides suggested layouts for forms, but the actual forms' design is a matter of local standards.

Derivation   A statement of where the information in the product was derived, e.g. from discussion with users, input from Project Management or Required System LDM.

Quality   One of the key innovations to SSADM with Version 4, this is the predefined criteria which the product must match if it is to be accepted. Before any activity begins, the team must know what level of quality it is to meet. This might be at the level of asking if every box on a form is completed, or checking the accuracy of data flows on a Data Flow Diagram, or specifying the cross-checks between two techniques or products. Again, the precise criteria are tailorable for each project; the important thing is that they are set *before* the activity which produces it is started.

External dependencies   A statement of any input to the product that comes from outside core SSADM (see Chapter 2). In addition, any outside body who will need that product, such as a user at a review, may be mentioned here. Many products will have no external dependencies.

References A list of references to any subsequent SSADM activity that uses these products as input.

WHEN

The SSADM dictionary defines the products, as above, to tell us about the *what* in SSADM. The Structural Model of the method tells us *when* each is to be produced. The Structural Model describes the hierarchy of activities, from module through stages to steps and tasks. Each of these activities is triggered by products from the Product Breakdown Structure, and, in its turn, produces others to feed back to Project Management. (See Chapter 2 for the SSADM Structural Model.)

HOW

The techniques of SSADM describe *how* the products are to be produced. The quality criteria are supported by the rigorous description of the techniques, and the tighter cross-validation that is now performed between the complementary techniques. Automated support, such as CASE, helps too in the generation of the products, and their validation. For some of the products that support the SSADM practitioner the CCTA has published a series of subject guides to illustrate the techniques concerned, and how they fit alongside the core method.

SUMMARY

SSADM now means Version 4 of the method. This version was introduced in 1990, after a series of pre-release briefings and seminars, aimed at encouraging its smooth absorption into the SSADM user community.

The changes are structural and procedural, having modified the shape of the method, and adapted the techniques.

The emphasis is now on products instead of activities. The whole method can be summarized in three threads that are mutually complementary:

● product-oriented: *what*.
● flexible structure: *when*.
● three views of model: *how*.

2. Shape of SSADM

2.1 Aims of chapter

In this chapter you will learn about:
- the place of SSADM in the project lifecycle,
- the SSADM Structural Model,
- the information highway,
- project activities and
- Project Procedures and CCTA subject guides.

2.2 SSADM in the project lifecycle

An organization that uses SSADM in the analysis and design of its IT projects is assumed to have reached a level of maturity in its use of IT. If it is still being tentative about computerization, or alternatively is allowing a number of uncoordinated, ad hoc systems to be built, it is probably not ready to make the commitment of resources and effort needed to implement IT successfully. It is therefore probably not ready to make the commitment necessary to practise SSADM.

Commitment to SSADM

The analysis and design stages of an IT project represent the Full Study, but by no means represent all of the project effort. If these two activities are poorly executed, the resulting system will be inefficient, ineffective and a liability. However good the programmers, if they are given inadequate specifications, they will not produce code to suit the Users' Requirements.

A prerequisite, therefore, for the use of SSADM is a commitment to providing the resources and organizational structure to allow SSADM to be used properly. By 'properly', I mean in such a way that use of the method produces a proper specification of a system to meet its Users' Requirements, that it demands minimal maintenance, that it is flexible enough to incorporate later requirements and is portable between environments. Figure 2.1 shows the nature of the project structure in such an organization.

The three levels reflects the three threads that were described in Chapter 1: the what, the how and the when.

The *board* level, where strategy is examined and agreed, defines the what. In other words, it describes which areas must be automated, or at least studied for change, and, in very broad terms, what the requirements are for each.

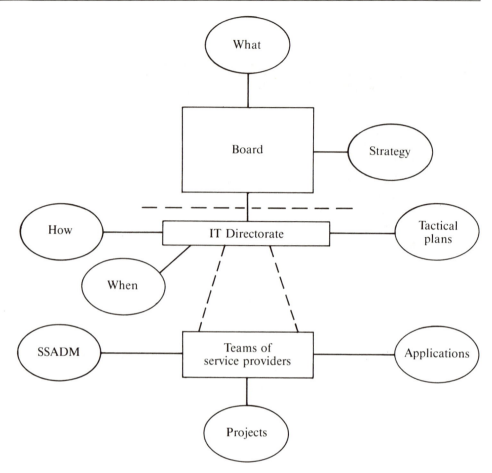

Figure 2.1

The *IT directorate*, at one level down, take this *what* and makes decisions as to *how* it is to be addressed. The directorate will also take the decision about *when* each is to be tackled.

At the bottom level, the *teams of service providers* carry out the SSADM and post-SSADM activities that make up the individual projects, that define *what* and *how* solutions are provided for the business requirements. The analysts, designers and programmers are all to be found here, as are other support teams that I shall mention below.

Progress of the project

The likely stages in an information systems (IS) project in an organization such as described in the previous section are as follows:

- an IS Strategic Study,

- an IS Tactical Study,
- a Feasibility Study,
- a Full Study, to produce a specification and
- a Development Project that produces and tests the physical system according to the specification.

Figure 2.2 shows the course of an IS project, and the place of SSADM inside it.

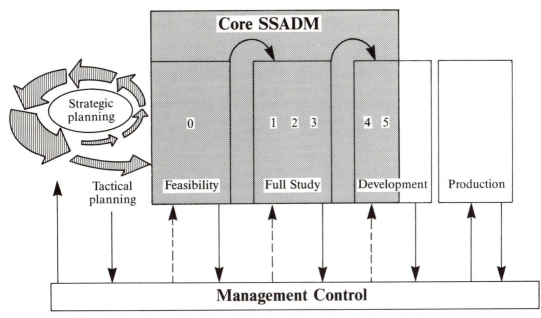

Figure 2.2

SSADM only covers the Feasibility and Full Studies. While particular SSADM techniques can be used during Strategic Study, another form of modelling the business and its broad requirements will be used then.

Feasibility

Whether or not a project is initiated as a result of an earlier Strategic Study, many will begin with Feasibility, to determine its course and obtain board commitment to it. Projects which are small and low risk are unlikely to need to undergo some form of feasibility study. The project team will receive their terms of reference from the findings of the study. The feasibility team will consider the proposed project from two principal views:

- *Technical feasibility*   Is the technology available to satisfy the requirements of the Users?
- *Business feasibility*   Can a sound business case be made for continuing the work? A Cost/Benefit Analysis must be made to assess the economic viability of the

proposals. Assuming that the proposed project is economically feasible, an impact analysis must be made to decide whether or not it is operationally feasible, i.e. will it work in this environment and culture?

If the answers to all these considerations are favourable, the Feasibility Report will determine the scope and direction of the project.

This is SSADM's first involvement with the project. Although SSADM has a Feasibility Module, with its defined stages and steps, and in this framework performs the tasks of Feasibility, there is more to Feasibility than simply the SSADM module. The CCTA's subject guide on Feasibility (Section 2.4) gives guidance on how SSADM is built into this study.

The next set of activities, the Full Study, is SSADM's principal contribution to the project and its success. The shape of that contribution is described in greater detail in Section 2.3. The end products of the Full Study are a detailed and optimized database design, ready to be implemented, and a set of programming specifications that are ready to turn into code, using either a 3GL or 4GL.

Post SSADM

The development work, which transforms these specifications into coded and tested software, is often regarded as a separate project. While results from it may well feed back into SSADM for revision, as an activity it is outside the scope of SSADM, as is the implementation of the system.

2.3   SSADM Structural Model

Structural Model requirements

SSADM encompasses a set of tools and techniques to be applied to the tasks of analysis and design. These tools and techniques are placed in a structure that prescribes when and how they are used. The aim of them all is to model the information requirements of the business user, and to specify the solutions in sufficient detail for the development team to build them.

There are other techniques, apart from the standard ones for systems analysis, and the problem is how to integrate these with the SSADM techniques. The solution chosen is to have two streams of activity: the SSADM activities, in their allotted place, and a set of ancillary activities that feed in the required information to support SSADM.

The model

The first stream is called *core SSADM*, and comprises a hierarchy of activities: Module → stage → step → task. There are five Modules in this stream, ranging from Feasibility through to Physical Design. In each of these Modules there are one or more stages, with defined activities and products.

The Modules are: Feasibility Study, Requirements Analysis, Requirements Specification, Logical System Specification and Physical Design. Each will be described in

Chapters 3 to 7 respectively. In these chapters I shall describe in detail the stages and steps of each Module, and all tasks associated with them.

Projects are essentially linear, and so the Modules are sequenced to allow the natural project lifecycle to be followed. However, each Module is a self-contained set of activities, and must be managed as a discrete project. It is possible that in an IT project, each Module will go to a separate contract, so that a different supplier may be employed on each. This indicates the importance of ensuring that the end of every activity is a set of products, to the required quality, so that another supplier may then take them as a starting point to continue the project.

Each Module will have its own set of plans, timescales, controls and monitoring procedures. In addition, each Module will have its own set of actors, who may not be the same through the whole project. It is highly unlikely, for example, that business analysts who model the requirements and required system will also produce an optimized physical data base design. Definition of the actors who will complete the tasks are part of the input to each Module.

The other stream is known as Project Procedures, and interfaces with core SSADM via Project Management activities. I shall say more about these procedures in Section 2.4. Chapter 21 describes them in more detail.

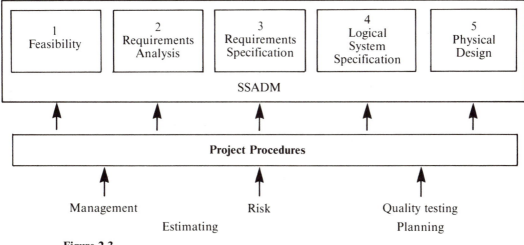

Figure 2.3

Figure 2.3 illustrates this relationship between the two streams.

Core Modules

The five Modules are defined in terms of the inputs and products. Inside each Module the decomposed activities are defined, also in terms of their inputs and products. Figures 2.4 to 2.8 show the five Modules, with the inputs, products and stages.

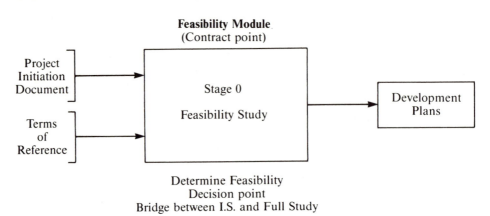

Feasibility Module
(Contract point)

Project Initiation Document

Terms of Reference

Stage 0

Feasibility Study

Development Plans

Determine Feasibility
Decision point
Bridge between I.S. and Full Study

Figure 2.4

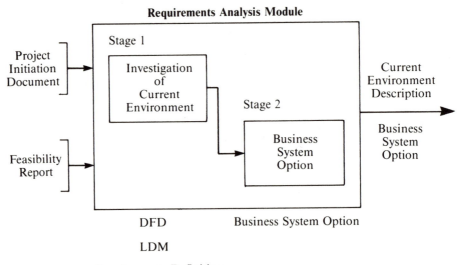

Requirements Analysis Module

Stage 1

Project Initiation Document

Investigation of Current Environment

Stage 2

Feasibility Report

Business System Option

Current Environment Description

Business System Option

DFD   Business System Option

LDM

Requirements Definition

Figure 2.5

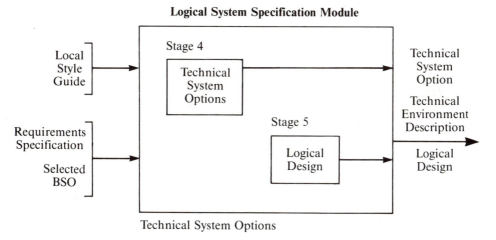

Requirements Specification Module

Current Environment Description →

Requirements Catalogue →

BSO →

Stage 3

Requirements Definition

→ Requirements Specification

→ Prototyping Report

Menu and Command Structures

Function Definition → RDA
→ Dialogue Design → Specification Prototyping
→ Event Entity Modelling

Figure 2.6

Logical System Specification Module

Local Style Guide →

Requirements Specification

Selected BSO →

Stage 4

Technical System Options

Stage 5

Logical Design

→ Technical System Option

Technical Environment Description

→ Logical Design

Technical System Options

Logical Database Process Design

Figure 2.7

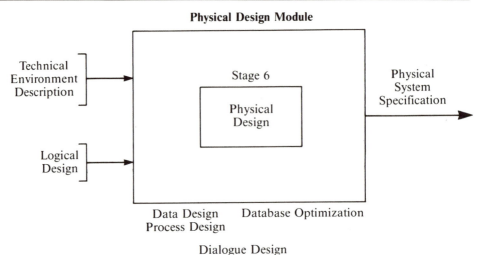

Figure 2.8

2.4   Support activities

I stated above that a second stream of activities, known as Project Procedures, supported the core SSADM activities. SSADM has been designed so that the place of such procedures in the project are well defined, and that all ancillary management plans and controls have their defined interface with the SSADM activities.

Before describing briefly the procedures and their place in the project, I shall describe the activity description standards and notation.

Activity interface

'Activity' covers the whole range of core SSADM, from the method, Modules, stages and steps to the contributory tasks that make up each step. Every activity is defined by a particular notation that specifies what happens, what it achieves, and how it is managed. SSADM prescribes the general format of these Activity Descriptions, but each project tailors them, or at least produces project-specific definitions for each. Figure 2.9 portrays the Activity Model that applies to every activity, large or small, undertaken in a project.

The activity itself is depicted by the striped box in the bottom half of the diagram. The Project Board, or whatever body is governing the project, is represented by the box *plan, monitor and control*. Inside this box, too, are implied all the Project Procedures that are described in Section 2.4.

The line marked *Information highway*, between Activity and Project Board represents the project manager. The manager is the buffer between the board and the practitioner team performing the activity, and is responsible for passing plans and constraints to the team, and submitting progress and other reports to the Project board. The end of the activity is the submission of actual products to the information highway at the conclusion.

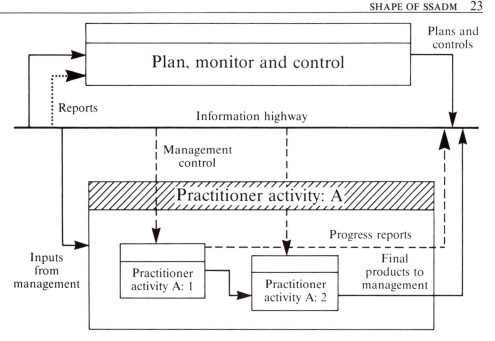

Figure 2.9

There are three kinds of flow represented on the diagrams:
1. products,
2. progress reports and
3. control flows/management authorization.

The actual labelling of these flows is the responsibility of Project management before the start of each activity.

Figure 2.10 uses the above notation to describe SSADM.

The Activity Description that accompanies the diagram describes the following features:
- objective,
- summary,
- participants,
- preconditions:
  —management authorization,
  —inputs,
  —references,
- products,
- techniques, and
- tasks.

Project Procedures

The philosophy behind the design of SSADM is that it can interface with other,

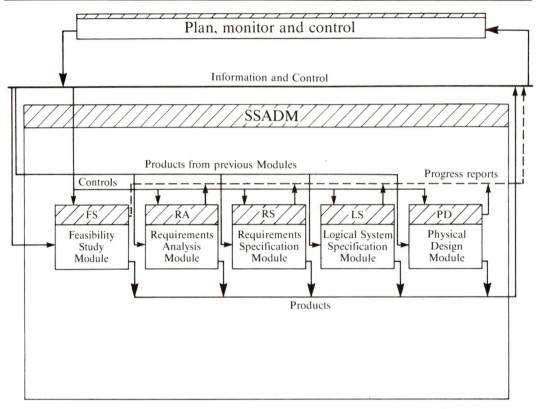

Figure 2.10

ancillary, activities as part of its normal planning. Any product which is produced outside core SSADM but must interface with it, is included on the hierarchical Product Breakdown Structure, in Fig. 1.2, and planned into the SSADM process.

All products passing between SSADM and these other activities do so via the information highway. All products that do appear must therefore have been planned for, and will appear as defined inputs or outputs to a SSADM activity at a predetermined point.

There are many activities that fall outside the scope of SSADM, yet are necessary for it to succeed. Principal ones are:

- Project Management,
- planning,
- estimating,
- Quality Control,
- Risk Assessment,
- Capacity Planning,
- testing,
- training,

- take-on,
- technical authoring and
- standards.

There are many others that a project will need to draw on, particularly if it has peculiar technical requirements such as distribution, or real-time facilities. Both of these can be handled by SSADM, but the method has to be tailored slightly to accommodate the requirements of the environment.

These activities support the SSADM practitioner; the information highway is responsible for passing the method's requirements to these procedures and these products into SSADM, it is essential that 'hooks' are provided for handling these interfaces.

Table 2.1 lists the above-named Project Procedures, whether they provide input to SSADM receive an output or both, and identifies with which part of SSADM they interface (i.e. the 'hook'). This hook may be a generic activity, such as Module or Stage, or it may be a specific point, such as Technical System Options.

Chapter 21 describes the activities and requirements of some of these procedures in more detail.

Table 2.1

Procedure	Interface	Level
Project Management Methods	I O	Modules, stages, via products/plans/structure
Standards	I	Modules, via installation standards
Quality Control	I O	Modules, stages, via products/descriptions
Capacity Planning	I O	Feasibility, BSO, TSO, via option descriptions
Risk Assessment	I O	Feasibility, BSO, TSO, Physical Design, via option descriptions, data descriptions, Function Descriptions
Take-on	O	TSO, Physical Design, via TSO products, and Physical Design products
Technical Authoring	O	BSO, TSO, Physical Design, via user Requirements, option description
Testing	O	TSO, via TSO/User Requirements
Training	O	Feasibility/BSO/TSO/Physical Design, via option description, design products

Subject guides

The emphasis on this second stream of activities interfacing with SSADM means that within the system suppliers' organization there must be both expertise in these procedures, and standards to apply to them. Many organizations taking up SSADM may not have standards for some of the activities, although if they build IS systems there must be a certain level of expertise.

As a support to SSADM, the CCTA, as part of the method documentation, have published a series of subject guides for the Project Procedures. They describe, but do not prescribe, particular ways of performing the activities. In some cases, e.g. risk assessment and management, they specify a proprietary method; in others, such as estimating and metrics, they describe principles and aids instead.

SUMMARY

SSADM is designed with two streams of activity operating: SSADM techniques, in their appointed place, and supporting techniques, Project Procedures, which provide inputs to SSADM, or receive SSADM products or progress reports.

The method is hierarchic, and made up of Modules, which are decomposed into stages then steps, which are made up of tasks, which make use of techniques.

There are five Modules:
1. Feasibility,
2. Requirements Analysis,
3. Requirements Specification,
4. Logical System Specification and
5. Physical Design

The output of each of these levels of activity is a set of products, predefined, and to specified quality criteria. The product orientation of the method makes Project Management more reliable, in that there is a measurable deliverable from every activity, rather than a bland statement of progress which, for whatever reason, may be misleading.

The Project Procedures are set in place by Project Management. The procedures are described in a set of subject guides, provided by the CCTA. Some of these guides describe proprietary methods in areas such as Project Management and Risk Assessment and Management, but the successful use of SSADM lies in employing the procedures according to a house standard, rather than in following a specific product.

3. Feasibility Study

3.1 Aims of chapter

The first part of this chapter describes the structure of the Feasibility Study Module. I shall describe the objectives and products of the stage that makes up this Module.

The second part is a breakdown of the procedures that are followed in the stage, detailing the activities carried out in all of the component steps.

3.2 Structure of the Feasibility Module

The Feasibility Study Module is carried out in one stage, Stage 0, at the outset of the project. It is triggered by the Project Initiation Document, which itself may be the result of an IS strategy document.

The IS strategy will already have indicated a direction for the project; as the project itself may not be undertaken for some time after publication of the strategy, the technical feasibility and business case must be re-examined, and a range of Business System Options and Technical System Options prepared.

The Feasibility Study is one of the checkpoints where either the project may be aborted, or a contract may be placed for a detailed study to be undertaken. Figure 3.1 shows the Feasibility Study Module's position at the start of the project, triggered by the Project Initiation Document.

Figure 3.2 illustrates the structure of Stage 0: Feasibility. All steps must be performed sequentially and in full.

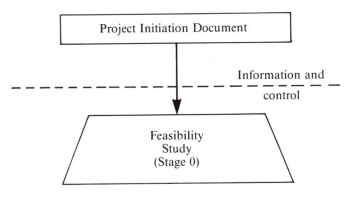

Figure 3.1

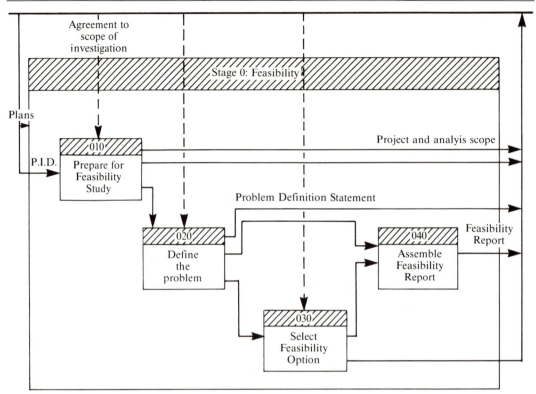

Figure 3.2

Feasibility

In this stage we carry out an assessment of a proposed IS system. We are trying to determine whether the system can, in fact, meet the business requirements it is intended to meet, and whether there is a sound business case for developing it. At the end of Feasibility, we should be able to say either that a full SSADM study should be undertaken, or that we should proceed in a different direction.

Most projects should begin with a Feasibility Study. The exceptions should be low-risk projects, which can be aborted, if necessary, at Business System Options. Feasibility should not be an expensive or time-consuming exercise: one team working for one to two months should be sufficient to produce the possible options and guidance for the way forward.

The study will involve the use of SSADM techniques to examine both the current system, and the requirements of the new system.

STAGE 0: OBJECTIVES

1.   To decide whether resources should be committed to a Full Study.

2. To decide whether to proceed in a different direction from that envisaged in the Project Initiation Document (PID) for the Feasibility Study.

STAGE 0: PRODUCTS
All Stage 0 products are passed to the information highway, or the information and control stream.
1. Feasibility Report:
    (a) introduction,
    (b) management/executive summary,
    (c) study approach,
    (d) existing business and IS support to the business,
    (e) future IS support required by the business,
    (f) proposed system,
    (g) options considered but rejected,
    (h) financial assessment,
    (i) project plans,
    (j) conclusions and recommendations and
    (k) technical annexes.

Steps
010   Prepare for the Feasibility Study
020   Define the problem
030   Identify Feasibility Options
040   Assemble Feasibility Report

3.3   Stage procedures

The remainder of this chapter describes the tasks performed in each of the steps that make up this stage.

Step 010   Prepare for the Feasibility Study
In this step we examine the PID and related documentation with the primary objective of agreeing the scope of the Feasibility Study, and planning the rest of the Module's work.

The requirements of the Feasibility Study will be reviewed with the terms of reference, description of the business environment and culture, the technical environment and business requirements/problems related to the project.

Before the project moves forward to Step 020 the Project Board must have agreed the scope and terms of reference, and resolved all questions and problems arising.

Techniques employed in Step 010 are:
● Data Flow Modelling
● Logical Data Modelling
● Requirements Definition
Figure 3.3 illustrates Step 010.

● Project Initiation Document
 ↓

Task	Description
10	Review the PID and relevant background material. Assess the scope and complexity of the proposed information system. Create a Context Diagram, current Level 1 DFD and overview LDS. Identify the base requirements from the PID and enter them in the Requirements Catalogue. Report errors or inconsistencies in the PID.
20	Identify the stakeholders for the business area concerned. Establish how they are to be involved, and brief the user representatives. Identify those areas to be investigated, and define the methods to be used. Agree the Feasibility Study scope with the project board.
30	Develop an Activity Network, Activity Descriptions, Product Flow Diagrams, Product Breakdown Structure and Product Descriptions. Agree these with the Project Board.

 ↓
● Context Diagram
● Current Physical DFD
● Overview LDS
● Requirements Catalogue
● Study plan
● Activity Network
● Activity Descriptions
● Product Flow Diagrams
● Product Breakdown Structure
● Product Descriptions

Figure 3.3

Step 020   Define the problem

In this step we carry out a more detailed investigation of the business and its information needs. We define the Users of the system, the new services wanted from the system, and any problems associated with the current services and operations.

    Techniques employed in Step 020 are:

● Data Flow Modelling
● Dialogue Design
● Logical Data Modelling
● Requirements Definition

Figure 3.4 illustrates Step 020.

- Context Diagram
- Current Physical DFM
- Overview LDS
- Requirements Catalogue

↓

Task	Description
10	Identify the activities and information in the area of study that are necessary for the business unit to meet its objectives. Draw a Level 1 DFD for the required environment. Amend the overview LDS to include entities and accesses in the required environment.
20	Investigate the current environment. If necessary, expand DFD to Level 2. Identify, with the user's help, those aspects of the current operations where improvement or change is required. Enter these in the Requirements Catalogue.
30	Define the Users of the new system in the User Catalogue.
40	Identify with the users new features in the required system (i.e. new processes, or functions). Record these in the Requirements Catalogue. Identify any non-functional requirements (i.e. service levels, response times, recovery and security). Record these in the Requirements Catalogue.
50	Prepare a problem definition statement to summarize the requirements, and to assess priorities for the business objectives.
60	Agree the problem definition statement with the Project Board.

↓

- Outline Current Environment Description
- Outline Required Environment Description
- Problem Definition Statement
- User Catalogue

Figure 3.4

Step 030   Identify Feasibility Options

In this step we develop a set of options for solutions to the problems identified and agreed. The options are presented to the Project Board, for the selection of one which will define the way forward. The selected option will define one or more projects to meet the requirements of the target business area.

The options combine both business option elements, to define the scope of the project(s), and technical options, to describe the physical environment in which it will operate.

After the board has made its selection, development plans are produced, in outline, for associated projects.

Techniques employed in Step 030 are:
- Business System Options
- Technical System Options
- Data Flow Modelling
- Logical Data Modelling

Figure 3.5 illustrates Step 030.

- Outline Current Environment Description
- Outline Required Environment Description
- Problem Definition Statement
- Requirements Catalogue
- User Catalogue

↓

Task	Description
10	Draw up a list of the minimum requirements for the new system, both functional and non-functional. All options must satisfy these basic requirements.
20	Define up to six Business System Options which will satisfy a range of requirements, from the basic minimum to all listed.
30	Define a list of outline Technical System Options to represent a full range of technical solutions. Each technical solution should satisfy the constraints and requirements of at least one Business System Option.
40	In consultation with the users, draw up a list of up to six composite Business and Technical Systems Options. Reduce these to a short list of about three.
50	Describe, in prose, each short-listed option. This description can be supported by DFDs and LDM, to illustrate the differences between them. They should be supported also by outline Cost/Benefit Analysis and impact analysis.
60	Identify the preferred option. Produce an outline development plan for each recommended project.
70	Present the short-listed options to the Project Board and other User audiences. Assist in the selection of the one chosen option, with explanation and clarification of the implications of each. Record any decisions made, with reasons.
80	Develop an Action Plan for the selected project(s). This should include a description of the technical approach and outline development plans.

↓

- Feasibility Options
- Action Plans

Figure 3.5

Step 040   Assemble Feasibility Report

In this step we publish the Feasibility Report, having checked the accuracy and integrity of the products of the study.

Each SSADM product is subject to a Quality Control check, but that is built into the Project Procedures rather than into an SSADM step. Ensuring integrity and consistency between the products is part of core SSADM. That happens in this step, and the corresponding step of each module.

Each Product Description carries a list of quality criteria, which serve as guides to the checks carried out here.

Figure 3.6 illustrates Step 040.

- Action Plan
- Feasibility Options
- Outline Current Environment Description
- Outline Required Environment Description
- Problem Definition Statement
- Requirements Catalogue
- User Catalogue
  ↓

Task	Description
10	Check the completeness and consistency of the Feasibility Study Module by reviewing the products above. Amend the products as a result of reviews, if necessary or appropriate.
20	Assemble and publish the Feasibility Report documents.

  ↓
- Feasibility Report

Figure 3.6

SUMMARY

Feasibility is an activity that should be undertaken in all but low-risk projects. It is a decision point; the decisions possible include that to terminate the project.

SSADM techniques are employed during the study, but they are not the only techniques used.

The options presented to the Project Board must address several issues: business, organizational, technical, and financial.

SSADM techniques used in Stage 0 are:
- Data Flow Modelling
- Logical Data Modelling
- Requirements Definition
- Dialogue Design
- Business System Options
- Technical System Options

On completion of this Module, if the project board give their approval, we are ready to move into the first Module of SSADM proper: Requirements Analysis.

4. Requirements Analysis

4.1 Aims of chapter

The first part of this chapter describes the structure of the Requirements Analysis Module, as well as the objectives and products of the two stages that make up this Module.

The second part is a breakdown of the procedures that are followed in each stage, detailing the activities carried out in all of the component steps.

4.2 Structure of Requirements Analysis

Requirements Analysis is carried out in two stages: investigation of Current System and Business System Options. Figure 4.1 illustrates the structure of the Requirements Analysis Module.

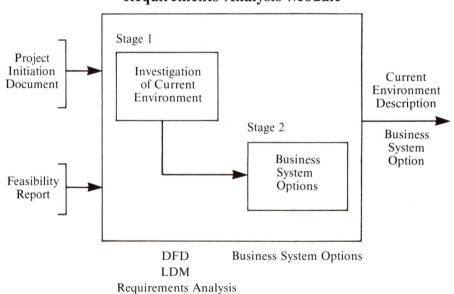

Requirements Analysis Module

Figure 4.1

Investigation of current environment

This stage may be the start of the project, triggered by a Project Initiation Document from the Project Board, or it may follow a Feasibility Study, and be triggered by the Feasibility Report.

However the stage—and hence the analysis and design project—is initiated, it has five objectives to fulfil.

STAGE 1: OBJECTIVES

1. To confirm that the project has the correct brief from management.
2. To prepare the initial task lists and resource estimates for the work.
3. To establish a clear statement of user requirements from the system, both functional and non-functional.
4. To establish roles for the project, especially the position and responsibilities of the users.
5. To model the procedures and data structures for those areas considered for computerization.

It is apparent that the first four objectives are more appropriate to the Project Management function, and the fifth is the one that corresponds to the traditional analysis stage of a project. The stage, certainly, is a planning stage, with the conduct of the project and the agreement of the project brief having equal importance to the investigation of the business environment.

STAGE 1: PRODUCTS

1. Passed to information and control:
    (a) Activity Descriptions,
    (b) Activity Network,
    (c) Product Breakdown Structure,
    (d) Product Descriptions,
    (e) Product Flow Diagram.
2. Passed to Stage 2:
    (a) Requirements Catalogue,
    (b) Current Environment Description:
        (i) Context Diagram,
        (ii) EPDs,
        (iii) I/O Descriptions,
        (iv) Current Environment LDM,
        (v) Logical DFDs,
        (vi) Logical Data Store/Entity Cross Reference.

Steps

110   Establish analysis framework
120   Investigate and define requirements
130   Investigate current processing
140   Investigate current data

150   Derive logical view of current services
160   Assemble investigation results

Business System Options

This Stage takes as its input the Current Environment Description, as described above, and the Project Initiation Document.

STAGE 2: OBJECTIVES

The objective of the stage is to select a solution to the problems and requirements of the given business environment. The scope of this particular project will finally be settled. Interfaces to other projects and other business areas will be agreed during this stage.

The solution will be selected from a number put up for consideration by the analysts. The options will consist in part of SSADM products to describe the information processing, and in part of financial, business and risk assessments to provide a sound business reason for the selection.

STAGE 2: PRODUCTS

Passed to Information and control:
(a) Business System Options, and
(b) Selected Business System Option.

Steps
210   Define Business System Options
220   Select Business System Options

Stage interfaces

As can be seen from this stage description, each stage and each module interfaces to Information and control as well as to the next module. Figure 4.2 illustrates these interfaces for the Requirements Analysis Module.

4.3   Stage procedures

As described in Chapter 2, each SSADM stage is broken down into steps, and each step is made up of tasks. The remainder of this chapter enumerates the steps and tasks performed in Requirements Analysis. The techniques performed in the tasks are described later in the book (Chapters 8–13). Standard SSADM forms used in these techniques will be described/illustrated in the relevant section of those chapters.

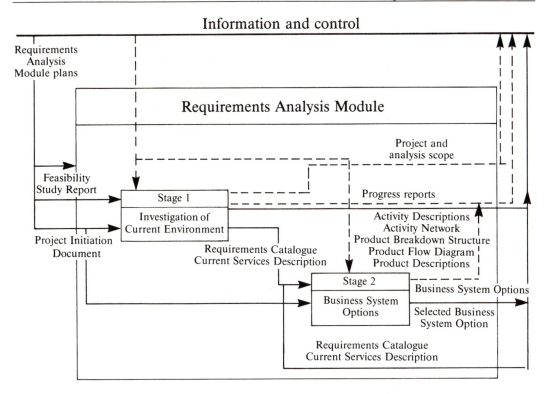

Figure 4.2

Investigation of Current Environment

Step 110   Establish analysis framework

This step is a concern of Project Management rather than IT analysis, or business analysis.

The Project Initiation Document, containing terms of reference, is examined, together with the results of any earlier study, whether SSADM Feasibility, or other. The inputs to this step will be reviewed to ensure that the project brief is still in line with the business objectives, and that both are consistent with the findings of earlier studies. Should any significant problems be identified here, the Project Board must be consulted before leaving this step.

At the end of this step the necessary project activities and products will be identified. They will form the basis for the Project Plans.

Techniques employed in Step 110 are:

- Data Flow Diagramming
- Logical Data Modelling
- Requirements Definition

Figure 4.3 describes Step 110.

- Feasibility Study Report, or
- Report from previous study
- Project Initiation Document (PID)
  ↓

Task	Description
10	Review the PID and the outputs from previous studies. Review Context Diagrams, Current Physical DFD and Over LDS, from earlier study. Make any changes needed to have them reflect the current environment. Identify specific system requirements from previous studies, and enter them in the Requirements Catalogue. Report any errors or inconsistencies in the input documents and wait until they are resolved.
20	Identify the target Users for the system, and establish their role in the analysis. Brief the User representatives. Identify the areas for investigation, and define the means of investigation (interview, questionnaire, document sampling, etc.) Agree the scope of the project and analysis with the Project Board.
30	Develop an Activity Network, Activity Descriptions, Product Flow Diagrams, Product Breakdown Structure and Product Descriptions for those SSADM parts of the project. Agree these items with the Project Board.

↓
- Activity Descriptions
- Activity Network
- Context Diagram
- Current Physical DFD
- Overview Logical Data Structure
- Product Flow Diagrams
- Product Descriptions
- Requirements Catalogue

Figure 4.3

Step 120   Investigate and define requirements
In this step we identify the problems in the current environment that are to be resolved, and also identify any additional services that the new system must provide.

The Requirements Catalogue (initiated in the previous step) is expanded in the light of more detailed investigation.

Requirements, under the twin headings of new requirements and current problems, are identified in broad terms only, and certainly no solutions are identified yet.

The technique employed in Step 120 is:
- Requirements Definition

Figure 4.4 illustrates Step 120.

Step 130   Investigate Current Processing
Step 130 describes the information flows in the environment in the form of Data Flow Diagrams. This activity is carried out in parallel with Step 120 (investigate and define requirements) and Step 140 (investigate current data).

- Requirements Catalogue
- User Catalogue
  ↓

Task	Description
10	Investigate the operation of the current system. Record details (for later use) of environment details e.g. volumes, frequencies, hardware usage, etc.
20	Define the intended users of the new system in the User Catalogue.
30	Identify from discussion with the users those aspects of the current system that present an opportunity for improvement. Enter these as problems in the Requirements Catalogue.
40	Identify from discussion with the user those functions and data not yet provided by the current system. Enter these as requirements in the Requirements Catalogue.

↓

- Requirements Catalogue

Figure 4.4

The Level 1 DFD from Step 110 is expanded and taken down to Levels 2 and 3. The technique employed in Step 130 is:
- Data Flow Modelling

Figure 4.5 describes Step 130.

- Context Diagram
- Current Physical Data Flow Diagram
- Requirements Catalogue
- User Catalogue
  ↓

Task	Description
10	For each flow on the Level 1 DFD, draw a Document Flow Diagram.
20	Combine the Document Flow Diagrams into one network. Use this network to improve the Level 1 DFD. Resolve any discrepancies between the two diagrams by discussion with the User. Steps 10 and 20 are only required if environment is complex, and needs clarification.
30	Draw a Level 2 DFD for each Level 1 process. Draw Level 3 processes where appropriate.
40	Create Elementary Process Descriptions for each lowest level process.
50	Create an Input/Output Description for each lowest level data flow across the system boundary.
60	Identify with the users any shortcomings in the current processing and record them in the Requirements Catalogue.

↓

- Context Diagram
- Current Physical DFDs
- Input/Output Descriptions

Figure 4.5

Step 140   Investigate current data

This step describes the logical structure of the data used in the current system. By 'logical' I mean that the structure of the data is divorced from the physical way it is held, in, for example, card indexes, or price lists, or filing cabinets.

The step is carried out in parallel with Step 120 (investigate and define requirements). At this point, the data model reflects only the requirements for the current processing.

The model will be validated against the processing documented in the Elementary Process Descriptions.

The technique employed in Step 140 is:

● Logical Data Modelling

Figure 4.6 illustrates Step 140.

● Overview Logical Data Structure
● Requirements Catalogue
● User Catalogue
● Elementary Process Descriptions

↓

Task	Description
10	Create an outline LDM of the current system data.
20	Define the important attributes associated with each entity.
30	Informally validate the model against the Elementary Process Descriptions.
40	Identify, with the user's assistance, any deficiencies in the availability of data, and record it in the Requirements Catalogue.

↓

● Current LDM
● Updated Requirements Catalogue

Figure 4.6

Step 150   Derive logical view of current services

This step refines the products of Step 130, by taking the DFDs of the current system and revising them so that they reflect the business logic of the system, rather than the physical implementation.

By identifying only the logical aspects of the processing, some of the problems previously identified will be resolved, such as redundant processing and anomalous procedures and data storage. Other problems and requirements will remain, however, but will not be considered at this point.

The techniques employed in Step 150 are:

● Data Flow Diagramming
● Requirements Definition

Figure 4.7 illustrates Step 150.

- Context Diagram
- Current Environment LDM
- Current Physical DFD
- I/O Descriptions
- Requirements Catalogue
- User Catalogue
↓

Task	Description
10	Remove traces of physical considerations from Levels 2 and 3 DFDs.
20	Rationalize the data stores, and relate each data store to one or more entities on the LDM.
30	Rationalize the processes on the lowest level DFDs, and group them together to form a Level 1 Logical DFD. Amend the supporting documentation (EPDs and I/O Descriptions) to reflect the new diagrams.
40	Cross-validate EPDs against the LDM.
50	Update the Requirements Catalogue to reflect those solutions associated with the physical constraints identified in this step.

↓

- Current Environment LDM
- Logical Data Store/Entity Cross Reference
- I/O Descriptions
- Logical Data Flow Diagrams
- Requirements Catalogue
- User Catalogue

Figure 4.7

Step 160   Assemble investigation results

This step completes the investigation of the current environment. Its objective is to ensure the consistency and integrity of the products of Stage 1. Quality reviews are held for all the products developed during the stage.

Figure 4.8 illustrates Step 160.
- Current Environment LDM
- Logical DFD
- Logical Data Store/Entity, Cross-Reference
- EPDs (current logical)
- I/O Descriptions (current logical)
- Requirements Catalogue
- User Catalogue
↓

Task	Description
10	Carry out quality reviews of the input products above. Amend the products as necessary in the light of the reviews.

↓

- Current Services Description
- Requirements Catalogue
- User Catalogue

Figure 4.8

Business System Options

Step 210   Define Business System Options

This step addresses the identified requirements by producing a number of possible solutions to them. Each of these solutions will be different to a greater or lesser degree, depending on impact, functionality and cost/benefit.

The technique employed in Step 210 is:

● Business System Options

Figure 4.9 illustrates Step 210.

● Current Services Description
● Project Initiation Document
● Requirements Catalogue
● User Catalogue
↓

Task	Description
10	Draw up a list of the requirements, both functional and non-functional.
20	Define up to six possible business solutions to these requirements.
30	Discuss the options with the users, and cut them down to a short list of three.
40	Describe each short-listed BSO. The description will be text supported by LDM and DFD, illustrating the differences between them. The descriptions must also include an analysis of the organizational impact, and a Cost/Benefit Analysis.

↓
● Business System Options

Figure 4.9

Step 220   Select Business System Options

In this step we present the short list to the Project Board, and help them to select the one that will be developed. Should the selected option be a mix-and-match affair from those on offer, as often happens, the definition of the option must be amended to cover the new functionality and features.

The technique employed in Step 220 is:

● Business System Options.

Figure 4.10 illustrates Step 220.

● Business System Options
  ↓

Task	Description
10	Formally present the BSOs to the Project Board. They will make a decision, with your assistance, on the most appropriate solution for the specified requirements. Record any decision reached.
20	Complete a description of the selected BSO. This will form the basis for the specification developed in Stage 3.
	If the selected option is exactly as the one represented, then most of the work is done; if the option chosen is a hybrid of two or more presented, a new description should be developed. This will include DFDs, LDS, Cost/Benefit Analysis, impact analysis and narrative. Also to be included are reasons for the selection of that option and the rejection of the others.

  ↓
● Selected Business System Option

Figure 4.10

SUMMARY

In Requirements Analysis we have carried out an investigation into the workings of the current environment. Using the two views of the data we have gained an understanding of the functional workings of the system, the data model that currently supports it, and a list of the problems and requirements that need to be addressed.

Using these as our starting point, we have now produced a number of solutions to those problems and requirements, and presented them to the user. On the basis of cost/benefit, organizational impact, functionality or whatever criterion is most important, the user has made a selection from the options. This selection may very likely be a hybrid composed of two or more of the options presented. Whatever has been chosen by the user must now be developed in sufficient detail to form the basis for the design of the new system.

Techniques employed in this Module are:
● Data Flow Modelling
● Requirements Analysis
● Logical Data Modelling
● Business System Options

The products from these activities may now be carried forward into the next module—Requirements Specification.

5. Requirements Specification

5.1 Aims of chapter

This chapter describes the activities and products of the requirements specification module. The second part of the chapter enumerates the steps and tasks to be undertaken.

5.2 Structure of Requirements Specification

This Module develops the work carried out in Requirements Analysis. The Business System Option having been selected, we must now consolidate our understanding—and indeed, the User's understanding—of the user requirements (Fig. 5.1). Requirements are both functional, and non-functional (i.e. to do with service levels, security, recovery, etc.).

New techniques are introduced to help us define the required processing, and to define the data structure to support that processing.

Definition of Requirements

This module comprises one stage only, Stage 3: Definition of Requirements. The products from Stage 1, particularly the Data Flow Diagrams and the Logical Data Model, are reviewed and reworked in the light of the selected Business System Option. The data model is enhanced by Relational Analysis and Entity Life History

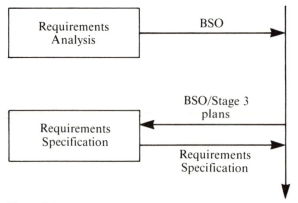

Figure 5.1

analysis, which specifies events in detail, and their effect on the modelled entities. The DFDs are translated into functions, and input/output (I/O) Structures are specified.

1. To provide for User management's approval a Requirements Specification Document for the development of a Logical System Specification.
2. To specify measurable acceptance criteria for subsequent design products.

All products from this stage are passed to information and control.
(a) Function Definitions,
(b) Requirements Catalogue,
(c) User Role/Function Matrix,
(d) Input/Output Structure,
(e) Prototyping Report,
(f) Required System Logical Data Model,
(g) Entity Life Histories,
(h) Effect Correspondence Diagrams and
(i) Enquiry Access Paths.
All of these products are combined in the document from Stage 3: Requirements Specification.

Steps
310   Define required system processing
320   Develop Required Data Model
330   Derive system functions
340   Enhance Required Data Model
350   Develop Specification Prototyping
360   Develop Process Specification
370   Confirm system objectives
380   Assemble Requirements Specification

5.3   Stage procedures

The eight steps listed above are not strictly sequential. Steps 310 and 320 are carried out in parallel. Step 310 feeds into 330, and Step 320 into 340. Step 350 may not begin until 310 and 330 are complete, while 330 and 340 are required to trigger 360. Figure 5.2 shows the structure of the stage.

Information and control

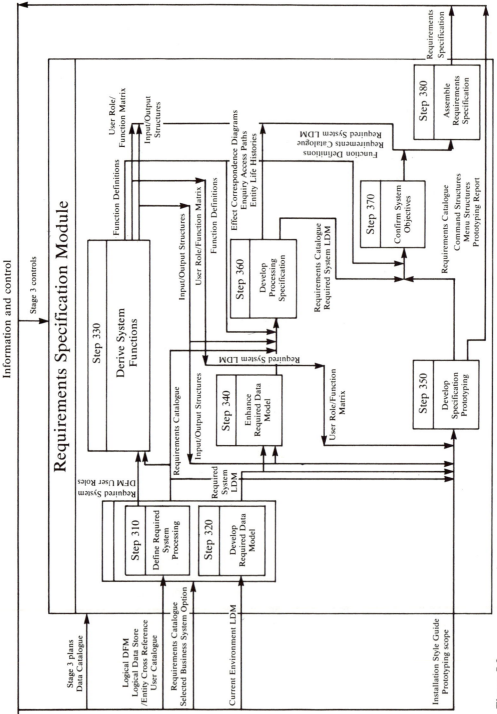

46

Figure 5.2

Definition of Requirements

Step 310   Define required system processing
Step 310 is carried out in parallel with Step 320. The Logical DFDs are fully amended to reflect the Selected BSO. The contents of all data flows across the system boundary must be defined in full at this point.

The various User Roles are identified here, for use during dialogue Design.

Techniques employed in Step 310 are:
- Data Flow Modelling
- Requirements Definition

Figure 5.3 illustrates Step 310.

- Logical Data Flow Diagrams
- Requirements Catalogue
- User Catalogue
- Selected BSO
- Logical Data Store/Entity Cross Reference
- Elementary Process Descriptions

↓

Task	Description
10	Redraw the Level 1 Current DFD to come into line with the selected BSO. Add any new processes identified in the BSO, and remove processes no longer inside the system boundary.
20	Amend the lower levels DFDs to support any new processing requirements. Update the Requirements Catalogue to include reference to the new or amended processes.
30	For each new lowest level process create an Elementary Process Description. Where appropriate amend existing EPDs. For each lowest level data flow crossing the system boundary, create new I/O Descriptions; amend existing descriptions where appropriate.
40	Validate the data stores on the required model against the Logical Data Model: make sure that each store consists of one or more entities. Ensure that the attributes listed against an entity are consistent with the data flows to and from the data store.
50	Define the User Roles in the required system, and ensure that these roles can be mapped on to the external entities on the required DFD.

↓

- User Roles
- Amended Requirements Catalogue
- I/O Description (required System)
- EPDs (required description)
- Logical Data Store/Entity Cross Reference

Figure 5.3

Step 320   Develop Required Data Model

This step is carried out in parallel with Step 310. The Logical Data Model of the current environment is amended to support the requirements defined in the Requirements Catalogue. All attributes of each entity will be defined in full.

Techniques employed in Step 320 are:

- Logical Data Modelling
- Requirements Analysis

Figure 5.4 illustrates Step 320.

- Current LDM
- Requirements Catalogue
- Selected BSO

↓

Task	Description
10	Review the Selected BSO and amend the Current LDM to take account of the additional requirements.
	Amend the Requirements Catalogue to reference the changes made for each requirement.
20	Validate the EPDs with the extended model.

↓

- Required System Data Model
- Amended Requirements Catalogue

Figure 5.4

Step 330   Derive system functions

In this step we identify functions for both updates and enquiries. For this we use the Required DFDs and the Requirements Catalogue.

Dialogues are specified—but not designed—and those critical to the system's success are identified.

Techniques employed in Step 330 are:

- Function Definition
- Dialogue Definition
- Requirements Definition

Figure 5.5 illustrates Step 330.

- Requirements Catalogue
- User Roles
- Required DFDs
- Required I/O Descriptions
- Required EPDs
- Logical Data Store/Entity Cross Reference

↓

Task	Description
10	With the Users, identify the update functions from the DFDs. Every lowest level DFD must have at least one function allocated to it. Identify the events contained within each function.
20	With the Users, identify all enquiry functions.
30	For each function, specify the I/O interface, using the I/O Descriptions from the DFD for the update functions. Consult with the Users for specification of the enquiry interfaces.
40	Cross reference the user roles and functions to identify the required system dialogues. Identify which dialogues are critical to the system.

↓

- I/O structures
- Function Definitions
- Requirements Catalogue
- User Role/Function Matrix

Figure 5.5

Step 340   Enhance Required Data Model

In this step, we validate and enhance the LDM by the use of Relational Data Analysis (RDA). The source of this analysis is the set of I/O Descriptions from Step 330. The normalized relations from this exercise are used to build submodels, which we then compare to the Logical Data Model.

    Techniques employed in Step 340 are:

- Relational Data Analysis
- Logical Data Modelling

Figure 5.6 illustrates Step 340.

- I/O Structures
- Required LDM

↓

Task	Description
10	Select those functions whose I/O Structures will be subject to RDA.
20	Carry out Relational Data Analysis to Third, or Boyce–Codd, Normal Form on the specified I/O Structures, and produce a set of normalized relations for each function.
30	Convert each set of normalized relations into a data submodel.
40	Compare the submodel with the corresponding part of the full data model. Resolve any discrepancies by consultation with the Users, and reference to the Requirements Catalogue.

↓

- Required System Logical Data Model

Figure 5.6

Step 350   Develop Specification Prototyping

In this step we develop prototype models of selected parts of the specification in order to validate that specification, and to ensure that the requirements are fully understood.

Techniques employed in Step 350 are:

- Specification Prototyping
- Requirements Definition
- Dialogue Definition

Figure 5.7 illustrates Step 350.

- I/O Structures
- Requirements Catalogue
- User Role/Function Matrix
- Installation Style Guide
- Prototyping scope

↓

Task	Description
10	Using the Prototyping scope (from Project Management), select the dialogues and reports to be prototyped.
20	Create prototypes of the dialogue Menu and Command Structures for the specified User Roles. Demonstrate the menu prototypes to the appropriate User Role. Modify and re-show as appropriate.
30	Identify the screen and report components that are to be prototyped. Create Prototype Pathways by combining them with the dialogue menus.
40	Implement the pathways.
50	Prepare for prototyping sessions with the users.
60	Carry out the prototyping sessions with the users.
70	Review the sessions, and report the results. Steps 40–70 may be repeated for each of the pathways being prototyped.
80	Review the results of the prototyping, and note any errors that have been identified in the Requirements Specification. Amend the Requirements Catalogue with details of the User interface requirements established during the prototyping sessions. Complete the report for management on the sessions.

↓

- Prototyping Report
- Menu and Command Structures
- Amended Requirements Catalogue

Figure 5.7

Step 360   Develop Process Specification

In Step 360 we define the processing requirements (for update only), expanding on the information gained from DFDs. We use a new tool, the Entity Life History analysis, a third view of the system data, in that it shows the effect of time on our entities. We identify the events that impact on our system, and model the effects for every entity.

The techniques employed in Step 360 are:

- Entity/Event Analysis
- Function Definition
- Requirements Definition

Figure 5.8 illustrates Step 360.

- Function Definitions
- I/O Structures
- Required LDM
- Requirements Catalogue

↓

Task	Description
10	For each entity on the LDM identify all events that update it, whether creation, modification or deletion. Work bottom-up from the LDS. The Function Definitions will have identified many of the events. For every event identified, create an Event Specification. For every new event identified, define the functions and amend the Function Definitions for redefined events.
20	Create a 'first-cut' Entity Life History for each entity on the LDS. For this first cut draw only a 'normal life'. After all have been drawn, include any 'parallel life' events.
30	Working top-down on the LDS, refine the ELHs to include abnormal death events, and any interaction between entities.
40	Add operations to the ELHs.
50	Amend the Requirements Catalogue with any new requirements identified in the course of ELH analysis
60	Draw an Effect Correspondence Diagram for each event.
70	Create an Enquiry Access Path for each enquiry function.

↓

- Entity Life Histories
- Effect Correspondence Diagram
- Event Specification
- Function Definition
- Requirements Catalogue
- Enquiry Access Paths

Figure 5.8

Step 370   Confirm system objectives

In this step, we re-examine the Requirements Catalogue to ensure that the identified requirements are met in the specification. Not only should the requirements be addressed, but measures must also be defined for assessing how well the requirements are met.

In addition to these tasks, we must examine the non-functional requirements to ensure that these are defined. Such requirements will be in the nature of operational requirements or constraints, or considerations such as audit requirements.

Techniques employed in Step 370 are:

- Function Definition
- Requirements Definition

Figure 5.9 illustrates Step 370.

- Function Definitions
- Required System LDM
- Requirements Catalogue

    ↓

Task	Description
10	Examine the Requirements Catalogue, to ensure that each functional requirement is fully defined. The definition should include such features as service levels and test criteria.
	After each definition is checked, ensure that each requirement is satisfied in the new specification.
20	Identify any non-functional requirements that are not yet defined. Ensure that all non-functional requirements are defined and met.

    ↓

- Function Definitions
- Required LDM   } amended
- Requirements Catalogue

Figure 5.9

Step 380   Assemble Requirements Specification

In this step, we assemble the documents for the Requirements Specification and subject them to a review. SSADM products are each given quality checks, but there is little attention given to the integrity of products passed between steps. Step 380 is one of the few points in the method where such integrity checks are made.

Once the reviews are complete, the formal document is published with all relevant end-of-module reports for management.

Figure 5.10 illustrates Step 380.

- Effect Correspondence Diagrams
- Entity Life Histories
- Event Specification
- Enquiry Access Paths
- Function Definitions
- Input/Output Structure
- Required LDM
- Requirements Catalogue
- User Role/Function Matrix

↓

Task	Description
10	Carry out reviews of the products that form the input to this step, ensuring completeness and consistency. Make any amendments to the documentation or products that the review process indicates.
20	Assemble and publish the Requirements Specification Document.

↓

- Requirements Specification

Figure 5.10

SUMMARY

In this module we have developed the specification of requirements, based on the Business System Option selected in the previous stage.

We have increased our knowledge of the environment by the use of Event-Entity Analysis, and Relational Data Analysis; we have broadened our understanding of the requirements of the new system by the use of prototyping with the user, and by defining fully the functions to be supported by the system.

Techniques employed in this Module are:

- Relational Data Analysis
- Entity Life History analysis
- Specification Prototyping
- Function Definition
- Requirements Definition

Having completed this Module, we are now ready to move forward to develop this specification in the first part of the design activities: Logical System Specification.

6. Logical System Specification

6.1 Aims of chapter

The first part of this chapter describes the structure of the Logical System Specification Module, as well as the objectives and products of the two stages that make up this module.

The second part is a breakdown of the procedures that are followed in each stage, and details the activities carried out in all of the component steps.

6.2 Structure of Logical System Specification

Logical System Specification is carried out in two stages: Stage 4—Technical System Options and Stage 5—Logical Design. The Requirements Specification and Business System Option from the two previous Modules are brought forward, and solutions found on two fronts: first, a Technical Option is selected, specifying the technical and development environments. As with the Business System Option, a number of options are produced and presented, and one is selected by the Users as that most appropriate to their needs.

After the Technical System Option is produced, we move into the design activities. At this point we are still looking at the project in logical rather than physical terms. Activities carried out here embrace Logical Design of update processes, Logical Design of enquiries and Logical Design of dialogues.

Figure 6.1 shows the composition of Logical System Specification. Figure 6.2 illustrates the interface between the three modules discussed so far and the Data and Control Highway.

Figure 6.3 presents the Activity Diagram for the Logical System Specification Module.

Technical System Options

In this stage we devise up to six technical options to implement our business solution, and present them to User management. They will make a choice—probably a hybrid of several features of different options. Very likely, the range of options will be much fewer than six, and will be constrained by the corporate IS strategy, the existing procurement policy or the findings of the Feasibility Study.

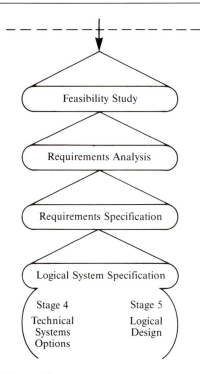

Figure 6.1

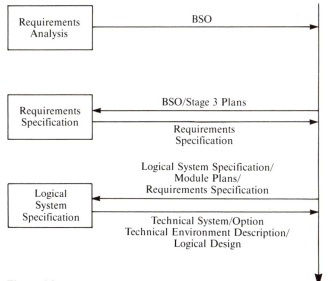

Figure 6.2

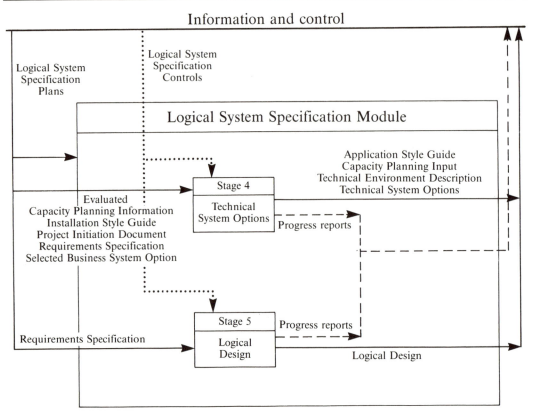

Figure 6.3

If a turnkey solution is nominated, then the analyst's description is necessarily limited to, for example, number and location of peripherals, volumes and required performance levels.

STAGE 4: OBJECTIVES
1. To identify and define ways of physically implementing the Requirements Specification.
2. To carry out a validation of the non-functional requirements in the light of the proposed technical environment.

STAGE 4: PRODUCTS
All Stage 4 products are passed to Information and control:
(a) progress reports,
(b) Technical System Option,
(c) Technical Environment Description and
(d) Application Style Guide.

Steps
410   Define Technical System Options
420   Select Technical System Option

Logical Design

This stage is carried out in parallel with Stage 4: there is no dependency between them, and so no requirement to wait for Technical System Options to be completed. Stage 5 sees the creation of logical dialogues with Menu Structures and Dialogue Control Tables, Update Processing Models, Report Formats, and Enquiry Processing Models, to produce the Logical Design of the required system.

STAGE 5: OBJECTIVES
1. To create the logical specification of the system defined by the BSO and Technical System Option.
2. To define the update and enquiry processing, and dialogues of the new systems, and to ensure their integrity.

STAGE 5: PRODUCTS
All Stage 5 products are passed to Information and control:
(a)   Logical System Specification.

Steps
510   Design User Dialogues
520   Define Update Processing Model
530   Define Enquiry Processing Model
540   Assemble Logical Design

6.3   Stage procedures

The remainder of this chapter enumerates the tasks performed in each of the steps that make up these stages. Although the stages are numbered 4 and 5, there is no dependency between them. The document Requirements Specification is input to each stage, and is not amended between the two. This implies that it is not necessary to wait for Stage 4 to be completed before beginning work on Logical Design. Since much of the work in Technical System Options may involve waiting for information from suppliers or technical/operations support staff, it is useful to be able to carry out parallel activities like this.

Technical System Options

Step 410   Define Technical System Options
In this step we identify several possible ways of implementing the Requirements Specification. We also confirm and validate the non-functional (service) requirements in the light of our specified technical environments.

We will probably be constrained in our choice of environments as a result of corporate IS strategy and/or Feasibility. Whatever constraints (or requirements) apply should be a part of the Requirements Catalogue.

The products of this step will be up to six documented options for presentation to management. Included in the documentation for each will be criteria such as cost/benefit and impact analysis to help management make an informed decision on the course to follow.

The technique employed in Step 410 is:
- Technical System Options

Figure 6.4 illustrates Step 410.

- Requirements Specification
- Selected Business System Option
- Project Initiation Document
- Evaluated Capacity Planning Information
    ↓

Task	Description
10	Examine all the input documents and identify the constraints that the options must meet.
20	Define a number, not greater than six, of technical solutions to the requirements.
30	Discuss these options with the User and discard any that they regard as non-starters. This should leave you with a short list of two or three.
40	Expand each surviving option into a full description, including for each: ● Technical Environment Description ● System Description
50	For each option, assess the capacity planning information. The service-level requirements must be met for each option, or else the variances must be described in the Technical Environment Description.
60	For each option, add: ● Impact Analysis ● Cost/Benefit Analysis ● Outline Development Plan

    ↓
- Technical System Options
- Technical Environment Descriptions

Figure 6.4

Step 402   Select Technical System Options

In this step, we present our shortlisted options to the Project Board, and help them make a selection. The decision will then be documented, and the context for Physical Design so defined.

There should be no surprises for the board at the presentation, as we will have discussed all the possible options with them before agreeing the short list. Even so, they may not choose one option as it stands, but 'mix and match' from the different ones on offer. If their choice is such a hybrid, then it too must be analysed, described and documented as the others were.

The technique employed in Step 420 is:

● Technical System Options

Figure 6.5 illustrates Step 420.

● Technical System Options
● Installation Style Guide
↓

Task	Description
10	Make a presentation to the board of each option. Help in the decision-making, and note the reasons for the decision.
20	Amend the selected option if appropriate. Develop the Technical Environment Description.
30	Ensure that the service level requirements will still be met. The tool for this is capacity planning.
40	Develop an Application Style Guide, based on the installation's own standards.

↓
● Selected Technical System Option
● Technical Environment Descriptions
● Application Style Guide

Figure 6.5

Step 510   Design User dialogues
In this step we design the logical dialogue for the system. Note, it is not the physical dialogue or physical screens that we are designing, but the logic of the exchange of data.
The techniques employed in Step 510 are:
● Dialogue Design
● Requirements Definition

Figure 6.6 illustrates Step 510.

● I/O Structures
● Function Definitions
● Requirements Catalogue
● User Role/Function Matrix
● Installation Style Guide
↓

Task	Description
10	Specify the syntax checks needed for input data.
20	Identify the logical grouping of elements in the dialogue.
30	Identify the navigation paths in each dialogue. Complete the Dialogue Control Table.
40	Define a menu set for each user role. Define valid control paths on completion of each dialogue.
50	Agree and define the 'help' requirements for each level of dialogue.

↓
● Dialogue Structures
● Menu and Command Structures
● Dialogue Control Tables
● Requirements Catalogue

Figure 6.6

Step 520   Define Update Processing Model

In this step we specify the logic of each database update required for an event. The required updates were defined in Stage 3, Definition of Requirements. Now we consolidate the entity updates into a single process structure for each event.

The ELHs from Stage 3 are completed with the addition of state indicators, and then the Effect Correspondence Diagrams are developed into a processing structure for that event. Semantic validation is defined for each event.

The techniques employed in Step 520 are:

- Function Definition
- Entity/Event Analysis

Figure 6.7 illustrates Step 520.

- Function Definitions
- Event Specification
- Entity Life Histories
- Effect Correspondence Diagrams
- I/O Structures
- Required System LDM
- Installation Style Guide

↓

Task	Description
10	Allocate state indicator values to the ELHs.
20	Convert the ECDs into a processing structure, one per event.
30	With reference to the ELHs, list the operations for each entity affected by the event.
40	Allocate the operations to the processing structure. Specify conditions governing selections and iterations.
50	Design formats for inputs and reports for the offline parts of update functions.

↓

- ELHs
- Update Processing Models
- Entity Descriptions
- Report Formats

Figure 6.7

Step 530   Define Enquiry Processing Model

In this step we complete the logical specification of all enquiry processing, as identified at Stage 3. We take the I/O Structures and Enquiry Access Paths already identified, and for each develop a single process structure. The structure is derived by merging an input structure (the Enquiry Access path) and an output structure (the I/O Descriptions). Operations and validations are added to this new structure.

The technique employed in Step 530 is:

- Function Definition

Figure 6.8 illustrates Step 530.

- I/O Structures
- Enquiry Access Paths
- Entity Life Histories
- Entity Descriptions
- Function Definitions
- Required System LDM
- Installation Style Guide

Task	Description
(Steps 10 to 40 will be carried out for each enquiry)	
10	Let the Enquiry Access Path be shown as a structure to represent the input data structure.
20	Let the I/O Description be shown as an output data structure.
30	Identify the points of correspondence between the two structures and merge them to form one process structure.
40	List the operations to be performed and allocate them to the new structure.
50	Design the formats for reports and inputs for the offline parts of the enquiry functions.

↓

- Enquiry Processing Models
- Report Formats

Figure 6.8

Step 540   Assemble Logical Design

This step completes and collates the Logical Design. It ensures the integrity and consistency of the respective products before passing them back to the project control organization.

Figure 6.9 illustrates Step 540.

- Dialogue Structures
- Dialogue Control Tables
- Entity Life Histories
- I/O Structures
- Function Definitions
- Menu and Command Structures
- Report Formats
- Required System LDM
- Enquiry Processing Models
- Update Processing Models
- Requirements Catalogue
- User Role/Function Matrix
  ↓

Task	Description
10	Examine the above products to ensure completeness and consistency. Amend the products if the reviews show this to be necessary.
20	Assemble the Logical System Specification, and publish it in accordance with the prevailing standards.

↓
- Logical System Specification

Figure 6.9

SUMMARY

Logical System Specification is the fourth Module in core SSADM. Its concerns are twofold:

1. To define the technical environment in which our new system must operate. This is achieved by selecting options for implementation in the form of hardware, software, development strategy and configuration strategy (e.g., centralized or distributed). Many of these decisions will be dictated by the corporate IS strategy, and/or the findings of the Feasibility Study.
2. To develop the products from Stage 3, Definition of Requirements, into a logical specification for the system. This involves specifying logical processes for updating the database, and for making enquiries of the database.
   The techniques applied in this Module are:

- Technical System Options
- Dialogue Design
- Logical Database Process Design
- Entity/Event Analysis
- Function Definition

On completion of this Module, we are ready to move into Physical Design, in Module 5.

7. Physical Design

7.1 Aims of chapter

The first part of this chapter describes the structure of the Physical Design module, as well as the objectives and products of the stage that makes up this module.

The second part is a breakdown of the procedures that are followed in Physical Design, and details the activities carried out in the component steps.

7.2 Structure of Physical Design

This module takes the Logical System Specification produced in Module 4 to create a physical database design, and program specifications to perform all the required functions, as seen in Fig. 7.1.

The stage comprises seven steps detailed below. Two new products created to achieve this aim are the Function Component Implementation Map (FCIM) and the Process Data Interface (PDI). These will be explained in detail later in this chapter, and in Chapters 16 and 17.

Physical Design

This stage looks at both aspects of design: data and processing. The data design is produced in steps: first of all we convert the Logical Data Model to a universal design that can be implemented regardless of DBMS. This first-cut design is converted to the

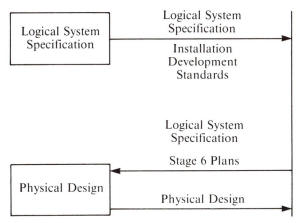

Figure 7.1

chosen DBMS, using the product-specific rules. The design is tuned to ensure that it meets the sizing and timing requirements of the new system. A strategy to implement this on the DBMS is drawn up, particularly where there cannot be a one-to-one mapping between the logical and physical data elements to be processed.

In conjunction with the data design, the Process Specifications are prepared according to an agreed style and format. A Process Data Interface is created to enable the processes to carry out the DB accesses where the one-to-one mapping is not possible, as indicated in the last paragraph.

STAGE 6: OBJECTIVES

1. To specify the physical data, processes, inputs and outputs, using the language and features of the chosen physical environment and incorporating installation standards.
2. The resulting design should provide everything needed to decide how application construction and introduction should be taken forward. User management will need to agree this with service providers in both systems construction and operations.

STAGE 6: PRODUCTS

All products from this stage are passed to Information and control:
(a) Physical Data Design (optimized),
(b) Dialogue Designs,
(c) Screen Designs,
(d) FCIM,
(e) Function Definitions,
(f) Process Data Interface,
(g) Required System Logical Data Model,
(h) Requirements Catalogue,
(i) Space Estimation and
(j) Timing Estimation.

Steps

610   Prepare for Physical Design
620   Create Physical Data Design
630   Create Function Component Implementation Map
640   Optimize Physical Data Design
650   Complete Function Specification
660   Consolidate Process Data Interface
670   Assemble Physical Design

7.3   Stage procedures

The seven steps listed above are not strictly sequential. Work on Step 610 can begin when the physical environment and implementation strategy have been chosen. All

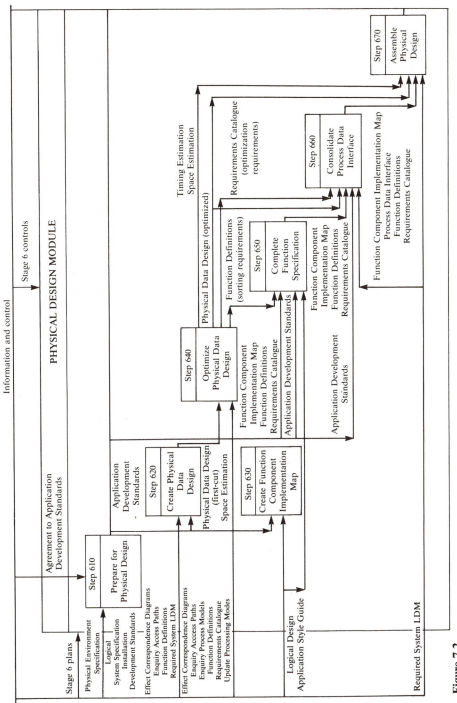

Figure 7.2

tasks must be completed before any subsequent steps are begun.

Part of Step 620—that part which specifies database access components—must wait until Step 660 is complete, although the other tasks can be carried out after 610. Steps 650, 660 and 670 will be carried out in sequence.

Figure 7.2 illustrates the structure of the stage.

Physical Design

Step 610   Prepare for Physical Design

This step allows the design team to gain an understanding of the physical environment prior to carrying out Physical Design. A physical environment classification scheme is used to categorize the physical environment. The scheme covers things such as data storage, performance and processing characteristics.

The characteristics and demand/constraints of the environment will clearly have an effect on the translation of the Logical Design.

Also in this step, the Application Development Standards are defined. The three main tasks here are to define the standards for the use of the physical processing system, to define the Program Specification formats, and to develop the Activity Descriptions for the Physical Design activities that are specific to the implementation environment.

The techniques employed are:

● Physical Data Design
● Physical Process Specification

Figure 7.3 illustrates Step 610.

● Installation Development Standards
● Logical System Specification
● Physical Environment Specification
↓

Task	Description
10	Complete the processing system classification for the processing environment. Complete the DBMS performance classification. Complete the DBMS data storage classification.
20	Design forms for the DBMS Space and Timing estimations.
30	Specify the standards for the use of the physical processing system and the DBMS facilities.
40	Specify the product-specific data design rules, unless already available.
50	Specify the naming standards for the application.
60	Develop the Program Specification and data design standards by defining a Product Breakdown Structure and Product Description for the Physical Design. Develop an Activity Network and Activity Descriptions for the remainder of Stage 6.
70	Initiate the preparation of manuals: user, operations and training. The production of these manuals will continue into the construction phases of the project, after SSADM is complete.
80	Agree the Physical Design Strategy with the Project Board.

↓
● Application Development Standards

Figure 7.3

Step 620   Create Physical Data Design

The aim of this step is to develop a Physical Data Design to implement the Required System LDM. An algorithm to convert the LDM into a first cut is followed. This follows general principles, and the first-cut design is non-specific.

This first design is then converted into a design specific to the DBMS to be used, using its own rules.

The technique employed in the step is:

● Physical Data Design

Figure 7.4 illustrates Step 620.

● Application Development Standards
● Effect Correspondence Diagrams
● Enquiry Access Paths
● Function Definitions
● Required System LDM
    ↓

Task	Description
10	Identify the features of the LDM that are required for Physical Data Design.
20	Identify the required entry points and distinguish those that are non-key.
30	Identify the roots of physical hierarchies
40	Identify the allowable physical groups for each non-root entry.
50	Apply the least dependent occurrence rule.
60	Determine the block size to be used.
70	Split the physical groups to fit the required block size.
80	Apply the product-specific data design rules for the target DBMS.

    ↓
● Physical Data Design (first cut)
● Space Estimation

Figure 7.4

Step 630   Create Function Component Implementation Map

This step is to specify those functions that are not included in the Logical Design, and also to describe those function components that can be specified non-procedurally.

The components not defined in the Logical System Specification (syntax error handling, physical I/O formats, physical dialogues, etc.) are specified. The Function Component Implementation Map (FCIM) identifies duplicate and common function components, and defines the relationship between all the function components.

Specification of database access components is not carried out until Step 660, as a part of the Process Data Interface.

Figure 7.5 illustrates Step 630.

- Application Development Standards
- Logical Design
        ↓

Task	Description
10	Identify and remove duplicate processing.
20	Identify and remove the specification of common processing.
30	Define success units.
40	Specify syntax error handing.
50	Specify controls and error handling.
60	Specify physical I/O formats.
70	Specify physical dialogue design.
80	Describe those function components that can be described non-procedurally.

        ↓

- Function Component Implementation Map
- Function Definitions
- Requirements Catalogue

N.B.   Tasks 30–80 are performed for each function.

Figure 7.5

Step 640   Optimize Physical Data Design

This step is to tune the first-cut product-specific data design to meet the space and timing objectives.

The Physical Design is validated against the performance levels defined in the Function Definitions and Requirements Catalogues. If the preset performance levels cannot be met, then the data design will be tuned.

The technique employed is:

- Physical Data Design

Figure 7.6 illustrates Step 640.

- Application Development Standards
- Effect Correspondence Diagrams
- Enquiry Access Paths
- Enquiry Process Models
- Function Definitions
- First-cut Physical Data Design
- Requirements Catalogue
- Space Estimation
- Update Processing Models
↓

Task	Description
10	Estimate the storage requirements.
	If necessary, restructure the data design to meet the requirements. Try to preserve a one-to-one logical-to-physical mapping.
20	Estimate the resource times of major functions.
	If the performance levels of these critical functions are not acceptable, alter the structure to achieve the desired access in time. Exploit the access mechanisms of the target DBMS to improve speed of access, but preserve the one-to-one mapping of logical-to-physical where possible.

↓

- Function Definitions
- Optimized Physical Data Design
- Requirements Catalogue
- Space Estimation
- Timing Estimation

Figure 7.6

Step 650   Complete Function Specification
In this step we specify those components of a function that cannot be specified procedurally. The results will be in sufficient detail for a programmer to code them, or for an application generator to produce the code.

This step is not an invariable step: it is only undertaken if components of the FCIM are to be specified non-procedurally. Where there are structure clashes, Specific Function Models will be produced. All function components needing procedural code will have program specifications written for them.

The technique employed in this step is:
- Physical Process Specification

Figure 7.7 illustrates Step 650.

- Application Development Standards
- FCIM
- Function Definitions (sorting requirements)
- Logical Design
- Requirements Catalogue

↓

Task	Description
10	Distinguish the logical processes in the function. If there is insufficient detail in the Function Definition, draw a Specific Function Model.
20	Combine the logical processes into a physical program or success unit.

↓

- FCIM
- Function Definitions
- Requirements Catalogue

Figure 7.7

Step 660   Consolidate Process Data Interface

In this step we complete the procedural specification, and examine the non-procedural implementation of the mapping between the Physical Design and the logical view of the data.

We compare the FCIM data access components with the optimized data design to identify any mismatches in the views of the data. The mismatches are resolved by identifying the keys and then the navigational sequence needed to pass the required view to the FCIM. This may be implemented by a coded module, procedurally, or with a non-procedural language, such as SQL, if this is used at the developers' site.

The set of PDI components is rationalized, any special maintenance or enhancement requirements are recorded. Design compromises are recorded in the Requirements Catalogue against the affected requirements.

The technique employed in this step is:

- Physical Process Specification

Figure 7.8 illustrates Step 660.

- Application Development Standards
- FCIM
- Functions Definitions
- Optimized Physical Data Design
- Required System LDM
- Requirements Catalogue

↓

Task	Description
10	Identify mismatches between a data access function component and the optimized data design. Step 640 will have identified some of these.
20	For each mismatch, identify the physical keys of the master and details to be accessed.
30	Starting at the top of each hierarchy affected, determine the physical access sequences to provide the FCIM access components. This will be done either with a procedural specification, or with the non-procedural language already identified.
40	Compare all the new processing components to identify duplicates.
50	Within the FCIM, record all the PDI elements that handle mismatches as being the subject of special maintenance and enhancement requirements. Note any which use special features in the physical environment, and those where low-level routines may be appropriate for performance reasons.
60	Annotate the Requirements Catalogue to show design decisions which limit the extent to which requirements have been met.

↓

- FCIM
- Function Definitions
- PDI
- Requirements Catalogue

Figure 7.8

Step 670   Assemble Physical Design

The Physical Design Module is complete at the end of this step, and the final Physical Design, the end of SSADM involvement in the project, is assembled and delivered.

Each product has its own quality criteria that have been built into the Product Description, and now we carry out a check for consistency between all these products.

Figure 7.9 illustrates Step 670.

- FCIM
- Function Definitions
- Optimized Physical Data Design
- PDI
- Required System LDM
- Requirements Catalogue
- Space Estimation
- Timing Estimation

↓

Task	Description
10	Check the completeness and consistency of the Physical Design Products by reviewing the inputs to this step, listed above.
20	Assemble and publish the Physical Design.

↓

- Physical Design

Figure 7.9

SUMMARY

In this module we have translated the Logical Specifications into a Physical Data Design and Program Specifications. The actual working system should now be ready to be written by a programmer using 3GL techniques, or fed into an application generator to produce the code.

Emphasis is placed, not on standards that are supplied by SSADM, but on those which belong to the local installation, or are specific to the vendor's own product, be it DBMS, 4GL, application generator, etc.

Much of the advice and guidance in the Physical Design Module is necessarily generic, and will be adapted to whatever physical implementation vehicle is selected. As always, it is up to the design team and Project Management to apply the guidance intelligently and appropriately.

PART TWO   SSADM techniques

8. Tools and techniques

8.1 Aims of chapter

This chapter introduces the analysis tools used in SSADM. For each one, a brief description of its purpose will be given, and an explanation of the notation. A fuller explanation of each use is given in the following chapters.

The last section of the chapter provides an overview environment of Super Systems (SS) plc, a training centre bookings function. This will be used throughout Part 2 to illustrate the techniques. It is not a full case study, and inconsistencies may be noticed in descriptions of different techniques. Its purpose is to present concrete illustrations of the use of SSADM techniques.

8.2 Introduction

I have said that SSADM describes the system from three different viewpoints: the functional model, the data model and the dynamic sequence model. The tools for describing these are, respectively, Data Flow Diagrams (DFDs), Logical Data Structures (LDSs) and Entity Life Histories (ELHs).

Each of these is described in detail in subsequent chapters with other related tools. My purpose here is simply to introduce them, and show the notation that each employs.

8.3 Data Flow Diagrams

Data Flow Diagrams are the core of most methods described as 'structured'. They come from the work of the Yourdon Corporation, and early structured analysis and design methods revolve around them. Yourdon, De Marco and Gane and Sarson's methods all use DFDs, each with its own notation and characteristics.

The virtues of DFDs are linked to their simplicity: the notation consists of just four elements, described below. First of all, I shall briefly explain why we need a simple notation like this.

After we have carried out our initial investigation of the current system, we need a way of describing its activities that we can show to the user to confirm or correct our understanding of the system. Traditionally, the two possible methods have been prose narratives or technical descriptions. Both of these are unsuitable.

Narrative is long-winded and prone to ambiguity. With the greatest care in the world, the analyst may still describe a section in terms that the user thinks he understands, but which is, in fact, inaccurate. Any mistakes that emerge in reviews may take considerable effort to correct, and in their correction, risk introducing further errors and ambiguities.

What is perhaps a greater drawback to narrative, however, is that it is necessarily long-winded and linear in structure. Thus it is harder to read and assimilate, and does not present a fair view of a system in which many things may be happening concurrently.

Technical descriptions are meaningful to those people who are involved in a given functional area, but less so to others. As the analyst is likely to investigate several functional areas in one study, there is a danger that several standards will dictate the contents of one document, and several different 'jargons' will be incorporated. This is not acceptable: it makes comprehension of the overall document harder, and amendment more fraught.

What is required is a relatively simple, comprehensive notation which gives an overall view of the organization, as well as appropriately detailed descriptions of activities. It should be easy and quick to draw, and easy to amend. Figure 8.1 shows a simple Data Flow Diagram describing the current physical system in a travel agency receiving a request for a booking for a given flight and hotel. There are just four elements in this diagram: *processes*, *data stores*, *data flows* and *external entities*. The meaning and structure of each is described in Fig. 8.2.

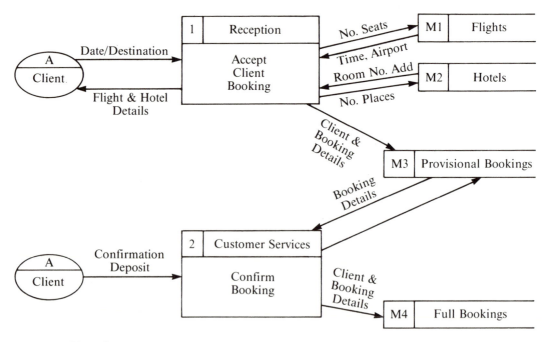

Figure 8.1

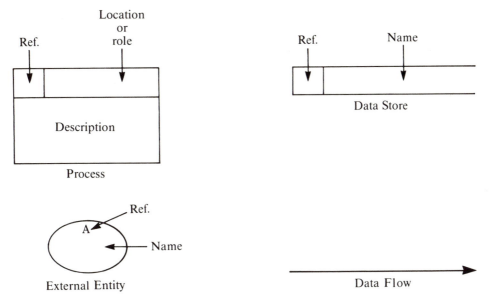

Figure 8.2

Processes

A process is an activity that receives data and carries out some form of trans-formation or manipulation before outputting it again. The activity may be carrying out calculations, creating a new document from information that triggered the process, or amending the document that flowed in. It is depicted by a box divided into three: the upper left position is given a number reference. This has no significance other than as a reference number; it does not imply priority or sequence. As a reference, for user communication, however, it is an important feature. The longer rectangle beside it names the location where the processing takes place; this may, on an overview DFD, be a broad term, Sales, Accounts, etc. As the DFDs become more detailed so do these descriptions.

The rest of the box describes what is happening in the process. The rule here is to keep the description as terse and meaningful as possible. Use an imperative verb and object, but make the verb specific. 'Process' and 'Update' are too vague and give little clue as to what is meant. 'Calculate', 'Add' and 'Validate' give a clearer picture of what is happening.

Data stores

A data store is a place where data comes to rest. It may be a filing cabinet, or an in-tray, a card index, a reference book or a computer file. Anywhere that data is stored and retrieved is called a data store.

The notation is simple: a long, open-ended rectangle, with a box at the left-hand end. The box is labelled with an alpha prefix and a number. The alpha is either D (for an automated data store) or M (for a manual/card data store). As with the processes, the number has no significance; it is purely a reference. The rectangle is labelled with a description of the contents of the data store.

If, for the sake of tidiness in the diagram, you wish to show the data store in more than one part of the diagram, draw a bar beside the left-hand box, as in Fig. 8.3. Each occurrence of the data store concerned will display that bar.

Figure 8.3

External entities

The third notation, a lozenge, represents an external entity. External entities are those bodies outside the system boundary which interact with the system. They may be external to the whole company, such as customers, Inland Revenue, Customs and Excise, or just external to the application area. Thus if we are modelling a sales office system, accounts and despatch areas would be shown as external entities. Each external entity communicates in some way with the system, so there is always a flow of data shown between a process in the system and an external entity.

The entities are labelled with a singular noun describing the role of the entity, e.g. Accounts, VAT Office, Credit Manager. Above the label will be an alphabetic character, again for reference purposes only.

As with data stores, it may be desirable for the sake of clarity to duplicate an external entity on the diagram, rather than have arrows from all points converging on one entity. If that is the case, put a small line along the top of the lozenge, as in Fig. 8.4.

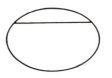

Figure 8.4

Data flows

Data flows represent any passage of data into the system, out of the system or between elements inside the system. It is represented by an arrow between the source and recipient of that data flow. In the world, it may take the form of a standard document with fixed content or a telephone call. It may be an enquiry, a functional document or a memo. Wherever traceable data is passed, it must be shown by an arrow. At the highest level DFD, one arrow may represent several data flows, which may be decomposed into the individual flows at the lower levels. Thus in our example, SS plc, a single data flow at the top level from Client to Bookings may read 'Booking'. As we expand and decompose the diagram, that one flow may decompose into 'Enquiry', 'Provisional Booking', 'Confirm Booking' and 'Cancel Booking'.

There are some validation rules about where data flows may or may not travel:

- Data stores may not be linked by data flows: flows must travel from one to another via a process.
- External entities may neither send nor receive data flows directly to or from a data store: they must communicate via a process.
- Data cannot be generated by a process, nor be swallowed by a process; documents may be swallowed or generated, but there must be output that is related directly to all inputs to the process.

8.4   Logical Data Structures

Logical Data Structures (LDSs) give us the second view of the system: the structure of the data. Chapter 10 on Logical Data Modelling describes the function and rationale for this technique. This chapter names the components and illustrates the notation.

An LDS diagram is made up of two elements: *entities* and *relationships*.

Entities

An entity is a thing about which we hold information in our system. It has the potential for more than one occurrence, and has a number of data items associated with it, to describe each occurrence. Typical entities in commercial systems would be Customer, Stock Item, Order and Invoice.

An entity is depicted by a 'soft' box (i.e. one with rounded corners), as in Fig. 8.5.

Figure 8.5

Relationships

A relationship is a logical business association between two entities. A customer, for example, will place an order for goods. Thus the three entities Customer, Order and Stock Item are related to each other. The relationships are shown by a straight line linking the entities.

Relationships in SSADM are normally in the form known as one-to-many, i.e. in a relationship one entity will occur once only, and the other may occur many times. The relationship joins the 'many' entity in what is known as a 'crow's foot'. However, in the description of the current environment, SSADM recognizes that one-to-one and many-to-many relationships may exist. These will be resolved into one-to-many before the required LDS is complete.

Figure 8.6 illustrates the Customer/Order/Stock Item relationship, showing the one-to-many relationships.

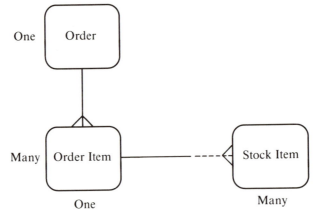

Figure 8.6

Types of relationship

There are several different types of relationship, such as *optional* relationships, (described in detail in Chapter 10), *exclusive* relationships and *recursive* relationships.

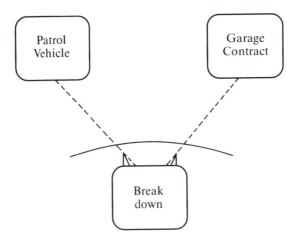

Figure 8.7

Figure 8.7 shows an exclusive relationship between three entities, Breakdown, Patrol Vehicle, Garage Contract. The environment is a motorway patrol system. The diagram shows that a breakdown may be attended by a Patrol Vehicle or a Garage

Contract vehicle, but not both. The arc is known as an *exclusion arc*, and denotes the either/or relationships.

Some entities may be related to other entities with the same key. The manufacturing bill of materials processing (BOMP) structure is the classic example of that. Figure 8.8 depicts BOMP.

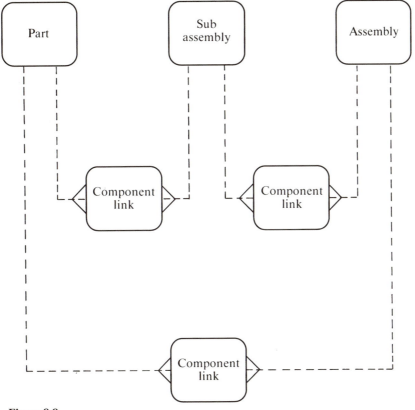

Figure 8.8

Each entity on the structure has the same key—Part No. The *link* entities will have a compound key of Part No./Part No., each Part No. reflecting one of the *master* keys. It can therefore be drawn more simply as a recursive structure, in Fig. 8.9.

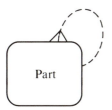

Figure 8.9

If the relationship is many-to-many, that is resolved by inserting a link entity with a double relationship, as in Fig. 8.10.

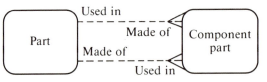

Figure 8.10

The entities in a relationship are either master or *detail*. The entity at the single end of a relationship is the master, and the entity at the many end is the detail.

8.5 Entity Life Histories

ELHs provide us with our third view of the system, the dynamic sequence, or time-based view. It shows the processing cycle of an entity, from creation to deletion. It models all the possible changes to the values of the attributes (data items) during its life, and the sequence in which the updates take place. That may be a little misleading: by 'sequence', I am not describing processing modes, but instead how business rules are implemented.

The business rules may stipulate, for example, that an entity occurrence may not be

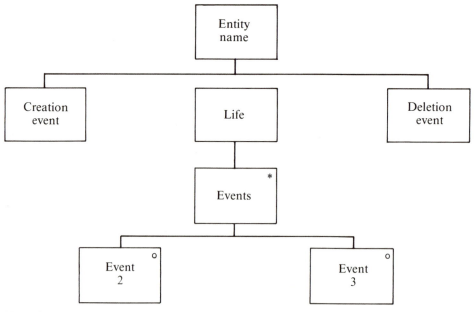

Figure 8.11

deleted from the system until certain conditions are met. The ELH stipulates the sequence in which updates to an entity must happen, and so is able to build in this particular rule.

Entity Life History uses the structure diagram notation (also known as Jackson-like structures, or Jackson-like diagrams). The structure diagram consists of three simple constructs: sequence, selection and iteration. Figure 8.11 illustrates a simple structure incorporating these constructs. The sequence is shown by the boxes, from left to right. The selection is shown by the boxes with a little circle in the top right corner. The iteration is depicted by a box with an asterisk in the top-right corner.

The single box at the top identifies the entity. The structure is read from left to right, top to bottom, starting with the creation (birth) of the entity and ending with the deletion (death). The boxes at the bottom of each branch are known as *effects*, and show the events in the real world that cause a change to be made to the entity.

A *selection* shows that at some recognized place in the Entity Life History, alternative events could affect the entity. Each possibility is catered for in the structure.

An *iteration* shows that an event may occur several times during the life of an entity, and each occurrence will cause an update to be made.

Another construct that can be used is that of a *parallel life*. this means that some updates to an entity can occur at any point in the life history; the timing of such events cannot be predicted, and so cannot be put into the sequence of events. Frequently, but not exclusively, this is associated with amendments to reference information on an entity. The notation to show a parallel life is a horizontal double bar below the 'life' structure box; from one end of the bar hangs the *normal life*, showing the sequence of events, and from the other hangs the 'parallel life'. Figure 8.12 on page 84 illustrates this.

8.6 Structure diagrams

SSADM makes extensive use of structure diagrams. They are found in Entity Life Histories, Effect Correspondence Diagrams, Enquiry Access Paths, Function Definition, Input/Output Structures, Dialogue Design and Logical Database Process Design (LDPD). Unfortunately, they are not always used in precisely the same ways, so in the early days of using SSADM care is needed to understand both syntax and semantics of whichever technique is being used.

In all cases the three basic constructs of sequence, selection and iteration are employed. Parallel lives are used only in ELHs. LDPD and Effect Correspondence Diagrams both use a further feature, the *correspondence arrow*.

The correspondence arrow makes a one-to-one correspondence between entities (on Effect Correspondence Diagrams) and data items in LDPD. In one-to-one, I include occasions when a single entity affected by an event corresponds to a set of related entities also affected. The correspondence here is one entity to one set of entities. If that correspondence always holds true, that is recognized as one-to-one.

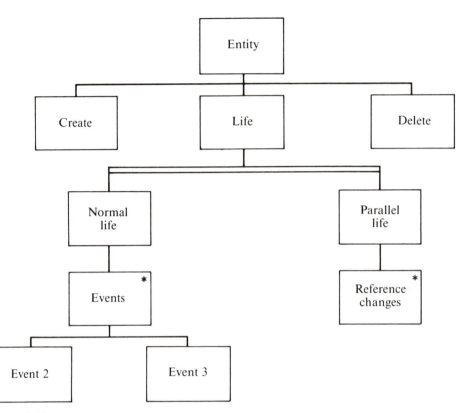

Figure 8.12

This situation is typical of master-to-detail entities: one master affected and a number of details of that master represents a one-to-one correspondence.

SUMMARY
SSADM makes use of three analysis tools to describe the system: Data Flow Diagrams for the function view, Logical Data Structures for the data view and Entity Life Histories for the time, or sequencing view.

The notation technique for Entity Life Histories, the structure diagram notation, is used in several different ways in SSADM, but the general principles for the notation are common to all.

Each of these techniques is supported by standard SSADM documentation, and is described in more detail in the rest of Part 2.

Figure 8.13 shows the place in SSADM where each technique is employed. The table shows the Modules rather than the stages which make use of the techniques.

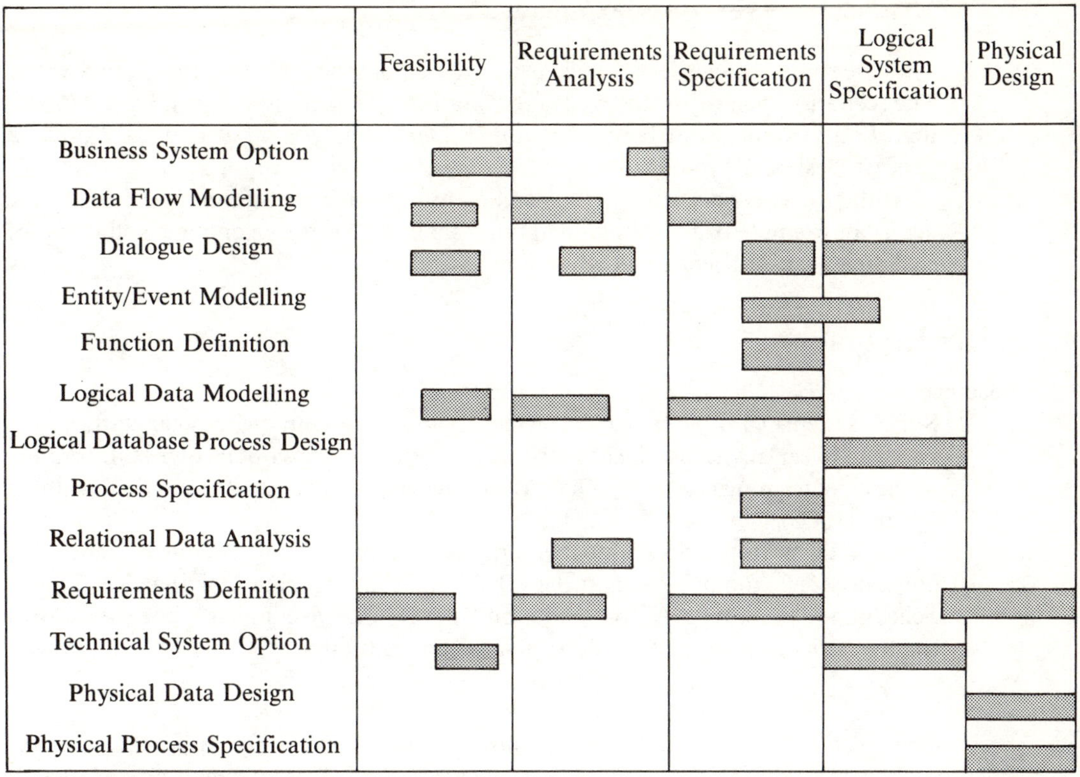

	Feasibility	Requirements Analysis	Requirements Specification	Logical System Specification	Physical Design
Business System Option	�		▪		
Data Flow Modelling	▪	▪	▪		
Dialogue Design	▪	▪	▪	▪	
Entity/Event Modelling			▪	▪	
Function Definition			▪		
Logical Data Modelling	▪	▪	▪		
Logical Database Process Design				▪	
Process Specification			▪		
Relational Data Analysis		▪	▪		
Requirements Definition	▪	▪	▪		▪
Technical System Option	▪			▪	
Physical Data Design					▪
Physical Process Specification					▪

Figure 8.13

The following chapters show how the different techniques interact.

CASE STUDY: 'SUPER SYSTEMS plc'

The case study outlined below will be used to demonstrate the techniques of SSADM. The scenario given forms the basis for Data Flow Modelling, Logical Data Modelling, Entity/Event Modelling, Relational Data Analysis and Logical Database Process Design.

As this study is only intended to demonstrate the different techniques, rather than act as an example of the whole method, I shall not produce complete solutions for each one, just sufficient to illustrate how it is used.

Super Systems plc

Environment

Super Systems (SS) plc is a nationwide organization employing 5000 staff, and is based at 70 centres across the country. The company carries out its own staff training at the chief training centre (CTC) in Milton Keynes and at 10 other regional training centres (RTC).

The CTC is responsible for scheduling and administering all courses across the company. The Admin section at the CTC is split into Course Maintenance, Course Scheduling, Bookings, Accommodation and Billing. Preliminary interviews with these functions revealed the following tasks and activities.

A   COURSE MAINTENANCE (CM)

This section is responsible for publishing the internal brochure. The Course Managers pass information about their courses to CM. This information is related to:

1. The creation of a new course title, with information about objectives, pre-requisites, duration, subjects covered, maximum number of students and level of staff who should attend (e.g. clerical staff, line management, storekeepers).
2. Amendment to published details—should any of the items in 1 above alter.
3. The removal from the portfolio of an existing course.

CM will incorporate these changes into a monthly amendment list, to be sent to all SS Training Officers.

B   COURSE SCHEDULING (CS)

CS publishes details for each run of the course titles, giving the location, start and end dates, and the name of the course contact (this will be either the Bookings Officer, or the Course Director).

A course that is in common demand will be scheduled to run on a regular basis: monthly, fortnightly, weekly, etc. Courses that have a limited take-up will only be scheduled to run when there are enough names on the waiting list to fill a run.

Each week CS examines the waiting lists for each course title; if there are sufficient to run the course, CS then examines the Training Schedule for free dates, and books a running of the course between three and six weeks hence. Facts that they need to establish are availability of premises and availability of appropriate teaching staff.

When such a course is scheduled the dates and delegate names are posted to the Nominating Manager, Course Manager, who will nominate a Course Director and teaching staff, Bookings and Accommodation.

In either event, no two runs of the same course will take place concurrently.

C   BOOKINGS (BK)

BK receives queries and requests for places from managers or Training Officers from SS's branches. If the request is for a course title with no scheduled runs, the names will be placed on a waiting list until a run of the course is scheduled. If there is a run scheduled, the delegate's name will be entered for that if there is still space. If the course is full, the delegate's name will be put on the waiting list for the next scheduled run.

BK receives details of all courses scheduled from CS when the dates are agreed. Ten working days before each scheduled course begins, BK sends joining instructions to each delegate's Nominating Manager or Training Officer, whoever is nominated to receive the instructions. Because of the volume of work that has grown recently, it sometimes happens that joining instructions are sent to the wrong delegates for a course.

If a delegate cancels before a course, BK tries to fill that place with someone from the same unit. If this is possible, the names are changed and no further action is taken. If not, the next name from the waiting list is substituted. If the delegate cancels less than two weeks before the course is due to begin, the Transfer Fee is forfeit. Otherwise it will be refunded, or credited for another Delegate from the same Nominating Manager.

If for any reason a Course Manager or CS cancels a course, BK is notified and has the tasks of informing all delegates' nominators.

A list of all attendees who have been booked for scheduled courses is sent to Billing at the end of each week. Arrangements for transfer fees can then be made.

D   BILLING (BI)

BI is responsible for sending requests for payment for each course run. Each booking is made by a Nominating Manager for one or more delegates. Payment (transfer fees, not cash) is made each month, and is for all delegates booked on to a course).

An arrangement may be made by which a special booking as in B above will be billed separately from the regular monthly bill, and paid at a special rate.

BI is responsible for both maintaining the price book for courses—a loose-leaf binder listing all course rates—and notifying all RTCs and Training Officers of any changes. They receive from Management—and distribute across the company—an amendment list of new prices twice a year.

Use of case study

For each of the exercises involving SS appropriate levels of detail will be given, if necessary, to enhance this outline. In this overview, for example, little mention is made of problems or requirements. Those chapters dealing with Requirements Definition and Function Definition will identify those problems and requirements pertaining to this system.

9. Data Flow Modelling

9.1 Aims of chapter

In this chapter you will learn:
- Where DFDs are used in SSADM.
- How to construct a current system DFD.
- How the Level 1 DFD can be decomposed to Levels 2 and 3 DFDs.
- How the Current Physical Model is logicalized.
- What documentation is needed to support a DFD.

9.2 Where Data Flow Modelling is used in SSADM

SSADM creates and amends DFDs in the following steps of analysis.

Step 110—Establish analysis framework   This step uses the results of any Feasibility Report to verify the Project Initiation Document. Any discrepancies between these and the current situation are amended by updating the DFDs.

Step 130—Investigate current processing   Here, we use the DFD as an aid to fact-finding and recording the results of the investigation. The overview DFD produced earlier is reviewed, and if necessary, extended in the light of the investigation. As the level of detail grows, so the DFD must be 'decomposed' to its lower levels—Level 2, or maybe even Level 3.

Step 150—Derive logical view of current services   The physical DFDs created above must be converted to give a *logical* view of the system, see Sec. 9.6. If there is no current environment the analysis begins with this diagram.

Step 210—Define Business System Options   To help the user select a business solution from a menu of possible solutions, the current view must be adapted to give a required view, see Sec. 9.7.

Step 220—Select Business System Option   Once the user has selected the option, Level 1 and Level 2 of the critical processes may be drawn to support it.

Step 310—Define Required System Processing   Using the Logical DFM, construct a full required DFM from the selected BSO and the Requirements Catalogue.

88

Step 330 ⎫   Derive System Scope   If any extra updates or data flows are identified
Step 360 ⎭   during these steps, the DFM is amended to reflect these.

Use in SSADM

In Chapter 8 I described the nature of DFDs and their notation in SSADM. In this chapter I shall look at how to construct a DFD for our case study, and how to convert this physical description of the current system into a diagram of the new, required system.

The term 'Data Flow Modelling' refers to a set of DFDs and the supporting documentation.

Inputs to DFM are as follows:
- Project Initiation Document
- Feasibility Report
- Current Physical DFM (for Logical DFM)
- Logical DFM
- Requirements Catalogue ⎫ for Required System DFM
- Selected Business System Option ⎭

Outputs from DFM are as follows:
- Data Catalogue
- Data Flow Model
- Logical Data Store/Entity Cross Reference

9.3   How to begin

The high-level DFD produced at Step 110 will identify the system boundry, the external entities, data flows across the boundary and the principal functional area/activities within the boundary. If the system is a particularly complex one, or the DFD technique is new to you so that you are not sure how to start, the following will give you a guide to producing that first high-level diagram.

The first thing is to establish a *context*, and to that end we can produce a *Context Diagram*. We then transform that context diagram into a Document Flow Diagram, and from there produce our Level 1 DFD.

Context Diagrams

If the high-level DFD is known as a Level 1 DFD the context diagram can be regarded as a Level 0 DFD. This can be drawn by following the steps below. To illustrate the techniques, I shall use the Super Systems case study.

Step 1   List the documents used in the system.
SS documents:
1. Bookings Form
2. Joining Instructions
3. Confirmation of Booking
4. Delegate List

5.   Newsletter
6.   Course Cancellation Notice
7.   Transfer Fee Request
8.   Payment/Advice
9.   Monthly Stats. Reports
10.   Delegate Cancellation
11.   Tutor Allocation
12.   Tutor Availability

Step 2   List all the sources and recipients of these documents.
SS sources/recipients:
1.   Bookings (BK)
2.   Billings (BI)
3.   Course Maintenance (CM)
4.   Course Scheduling (CS)
5.   Course Managers
6.   Tutors
7.   Delegates

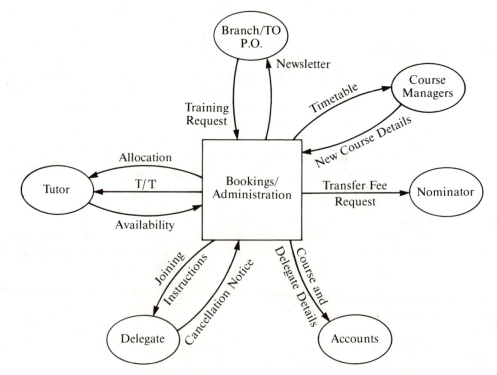

Figure 9.1

8. Branch Training Officers
9. Accounts Department
10. Branch Line Managers

Step 3 Draw a box representing the system and show the flow of documents from these sources and recipients, as in Fig. 9.1. Those areas which are known to be inside the system (here they are CM, CS, BK and BI) are hidden inside the box.

Document flows

From this Context Diagram, or Level 0 DFD, we can expand the diagram into a Document Flow Diagram by leaving out the 'black box', and showing the documents' path between the entities, as in Fig. 9.2.

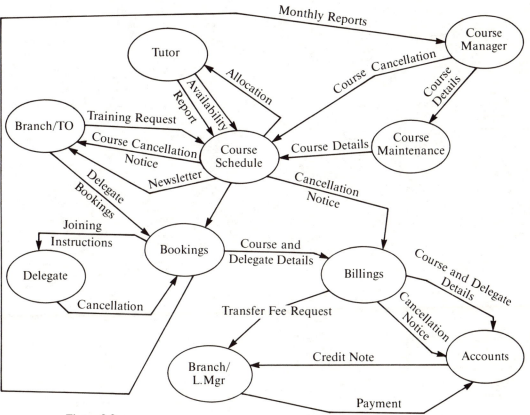

Figure 9.2

9.4   Developing the Physical Model

Top-level DFDs

Having drawn this document flow, we must, with the help of the Users, agree on the boundary. This can be marked by a dotted line, for ease of reference. Those entities outside the boundary are now known to be external entities. Those inside the boundary will be functional areas that act on the flows of data. The first thing to do at this stage, then, is to transform those entities within the boundary into processes, and label them accordingly, as in Fig. 9.3.

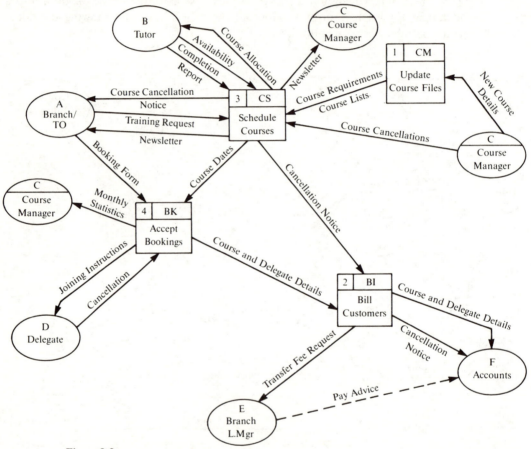

Figure 9.3

Our high-level DFD is not complete yet: we have not identified how many processes each of these functional areas performs, and what data stores are used. As we are describing a physical system, we must consider the types of data that are moving through the system. As a rule-of-thumb guide, we can consider the following types of data store:

1. Standing data, used for the day-to-day functioning of the system and kept up to date. Such items would be Course Brochures, Course Schedules, Hotel Lists.
2. Historical data that is required to be maintained for reference and enquiry purpose, but is no longer 'live'. Such items could be details of Course Presentations that are completed.
3. Temporary data stores, such as collections of delegates who require Joining Instructions to be sent out the next day, their details being batched together in the previous afternoon in readiness. Once the Joining Instructions are dispatched this collection of data will not be needed and the data store will no longer exist.
4. Extracted data that is retrieved from different sources for the purpose of preparing reports, statistics and so on.

What we must do at this point is to examine the data flows across the boundaries into the processes/functional areas on our 'first-cut' attempt, and see what actions are performed on them. Each process will probably require at least one data store. Our initial investigations will identify all such data stores and their access, so whatever is unclear at this point can be clarified then.

The labelling of the processes will help us to sharpen our understanding of what is happening in each area: the process should be identified by a clear, unambiguous verb with a direct object.

If all we can produce is 'Update Wait-list' we must break that down, now or later, into some such activity as 'Place new applicant on list', 'Remove applicant from list', and so on. In this way, we expand our initial attempt to give a clear, high-level description of what happens to our data flows in each section, and what data stores require to be accessed. Figure 9.4 gives us a view of the system after this process.

While this seems to be an accurate view of the organization's operations, it is not easy to read. As one of the virtues of the DFD is its use as a communications tool, we must try to tidy it up and make it intelligible.

There is no hard-and-fast rule for this, just several re-draws. A drawing tool, or even better, a CASE tool, would best serve us here. In terms of saving time, automation gives benefits, and in terms of preserving the integrity of the diagram during re-draws, CASE can ensure that data flows 'stick' to each object as it is moved about the screen, so that the logic of the re-draw is the same as the logic of the original, messy draft.

Two possible 'fixes' to help tidy the Level 1 diagram are combining related data flows, and related external entities. For example, we have two external entities relating to the branches: Branch/TO and Branch L Mgr. At Level 1 it is permissible to combine them to form one Branch as in Fig. 9.5.

On our DFD, we have three data flows coming from external entity Branch/TO into Process 4 Receive Application. At Level 1, these can be combined to form one composite data flow: Booking, as in Fig. 9.6.

Optimizations such as these will tidy up the Level 1 diagram, but must be decomposed again at Level 2, to ensure that we have modelled the activities in detail, and have not overlooked some important flow, or User Role. Having carried out our tidying up operations, Level 1 of SS plc now looks like Fig. 9.7 on page 96.

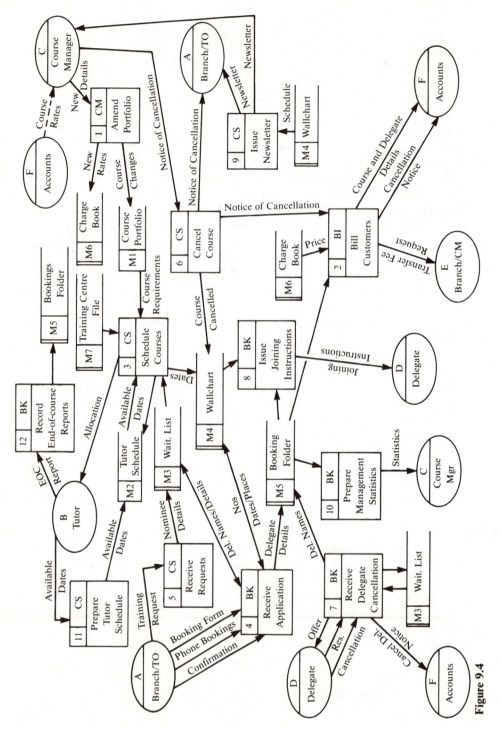

Figure 9.4

94

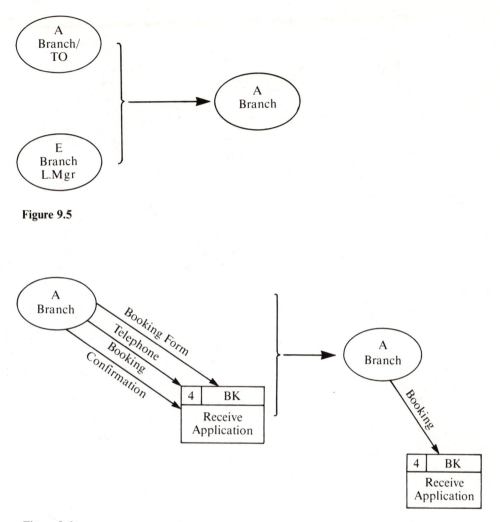

Figure 9.5

Figure 9.6

Validation of DFD

Before we move on from this initial high-level DFD, drawn up at Step 110, we must make sure that it is consistent. Below is a checklist of points to watch before moving on to the detailed investigation which will take us to the lower levels.

1. Has each process a strong imperative verb and an object?
2. Are data flows in related to data flows out? Data should not be swallowed up by a process, only transformed in some way. A data store is the only place data is allowed to rest. Similarly, data cannot be generated by a process. A document may be, but the data on the document comes from a data flow into that process.

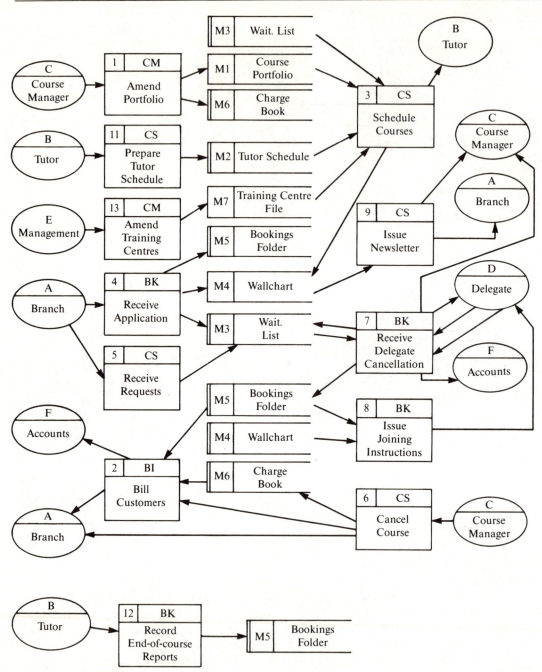

Figure 9.7

3. Can the flows be reduced? If a process is too busy, it can perhaps be broken down into two or more processes: six data flows in or out of a process should be sufficient.
4. Do all data stores have flows both in and out? A one-way data store is of little use, unless it is a temporary data store, or it is a reference file only. If the Current Physical DFD should identify such a data store, confirm with the User that you have correctly understood the procedures.
5. Are symbols correctly labelled and uniquely referenced?
6. Do all external entities communicate with a process? No entity should be allowed direct access to data, either to read or update it

When you are satisfied with these questions of DFD notation and logic, it is time to check with the User that the diagram accurately reflects the business system logic as well. If the User agrees that the diagram is an accurate portrayal of the work area, and you both agree on the 'domain of change' (i.e. the system boundary), then it is time to progress to the more detailed investigations of Task 130 that will lead to the decomposition of your overview.

Decomposition of top-level DFDs

The Level 1 DFD presents us with an overview of the system, a description that could come from a preliminary interview with departmental managers, perhaps. As we delve deeper into the operations of our system we find that we have to include rather more detail. This is done by examining the processes more closely and breaking each one down into other processes. The following algorithm will explain how this is done.

Step 1   Using standard SSADM form make each process box the system boundary. All data flows to or from that process are now flows across the lower level system boundary.

Step 2   Draw, outside the new boundary, the sources and recipients of these flows, as shown on the higher level DFD, be they external entities, data stores, or other processes. Ensure that they are labelled consistently with the higher level.

Step 3   Identify and draw the processes at the lower levels that act on these data flows. Number the sub-processes with a decimal extension of the higher level number, i.e. Level 1 Process 3 will break down to Processes 3.1, 3.2, 3.3, etc. Those process boxes that cannot be decomposed further, mark with an asterisk in the bottom right-hand corner.

Figure 9.8 illustrates these steps by looking at activity 'Receive Application' from the SS Level 1 DFD.

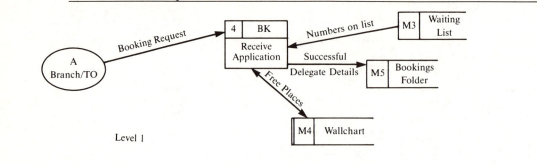

Level 1

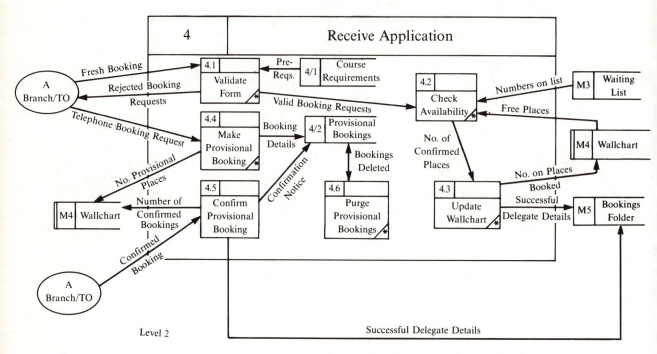

Level 2

Figure 9.8

Step 4   Carry out consistency checks, as before.

Step 5   Make sure that all lower level DFDs map on to the Level 1 diagram, by checking data flows.

Step 6   Review the lower levels with the User to ensure that you have depicted every activity actually performed in the system under investigation.

You may find that a single data flow on the top level is itself decomposed at Level 2. Thus, in Fig. 9.8, the Level 1 data flow 'Booking Request' is decomposed in

Processes 4.1 and 4.5 into Fresh Booking, and Confirmed Booking.

When you have taken your DFD as far as you can, the details must be recorded on an Elementary Process Description (EPD) using a concise and precise narrative. If more than four or five sentences are required, perhaps the process has still to be broken down to another level.

If a process involves making a decision, this is to be recorded on the EPD, *not* on the diagram. (Note: a common mistake made by people learning DFD technique is to treat the chart as a flowchart: this is *wrong*! Use a tool such as a decision table, or decision tree to describe decisions.) If the decisions are complex, i.e. more than five or six possible courses of action, the process will probably be too busy, and so capable of further decomposition.

9.5   Supporting documentation

A data dictionary, either paper or automatic, should be maintained at every stage of DFD production. The dictionary should contain:

- An *External Entity Description*   This describes all the external entities shown on the diagram. Included will be such details as the functions of the entity, and constraints on how it is to interface to the system.
- *Input/Output Description*   A list of all the data flows, the contents, and the start and end references for each flow crossing the system boundary.

It is important that this dictionary is maintained for the current system and for the required system. The details will grow with each iteration, of course; the first attempts are not expected to be more than a guide.

9.6   Transforming Physical DFDs into Logical DFDs

(Note: this happens in Step 150, after the Logical Data Model has been drawn (Logical Data Modelling is the subject of Chapter 10). This refers explicitly to the Logical Data Structure (LDS), and so to clarify the points being made there, the LDS for Super Systems plc will be drawn in Fig. 9.10).

Concept of 'rationalization'

Our DFD, with supporting documentation, has provided us with an accurate and graphical description of how the system currently operates. It is an account of the *physical* system; our next concern is to convert this to a *logical* view of the system. This means removing all signs of accidental or contingent procedures and leaving only those that are necessary to show *what* is happening.

The Logical DFD depicts *what* happens; the Physical DFD shows *how* it is carried out. Conceptually, this is one of the hardest aspects of DFDs, but the following example should clarify what I mean.

Assume that a clerk in an office requires a new piece of equipment, say a filing cabinet. The office procedures may dictate the following pattern:

1.   The clerk concerned will request the item from the office manager.

2. The manager will complete an equipment order form.
3. The form will pass to the Divisional Manager for approval.
4. The form will be photocopied, and the copy kept for Divisional records and audit purposes. The original will be put into a batch of other requests from the Division, to be forwarded to Purchases at the end of the week.
5. Purchases will complete a Purchase Order on Friday afternoon and despatch it to the suppliers, with all the other orders from Divisions.

This sequence is shown as a Physical DFD in Fig. 9.9a.

Logically, however, what happens is that an office requests an item, and notifies purchases who issue an order to the suppliers. This is depicted more simply as Fig. 9.9b.

This example shows that the object of this exercise is to remove the constraints of the physical work environment, leaving us with the simple data flows necessary to show *what* is happening.

To this end, we must identify processes performed for any of the following reasons:
1. *Procedural*   Such as batching forms, or sorting them.
2. *Machine/tool-related*   Such as typing or photocopying documents.
3. *Political*   To do with company policy, such as seeking approval from particular managers for a procedure.
4. *Geographical*   An organization spread over a wide area may need special procedures and data flows to meet ensuing problems.

To carry out this 'idealization', or logical conversion, we perform three tasks:
1. rationalize data stores,
2. rationalize processes and
3. group bottom-level processes into higher-level processes.

Rationalizing data stores

Data stores, in whatever form they are implemented, can be categorized as *main* stores or *transient* stores. Each of these should be approached differently.

MAIN DATA STORES

The data stores on our Logical DFD should have a correspondence with the entities on the Logical Data Structure. Figure 9.10 gives the Logical Data Structure for Super Systems plc.

The aim of this activity is to find a logical correspondence between the entities on the LDS and the data stores on the DFD. The end of the exercise will be to match all the attributes on the LDS with the data items in one and only one data store on the Logical DFD, i.e. such that no attribute will be found in more than one data store. Each logical data store will refer to one entity, or a group of related entities.

To achieve this, we must examine the data stores and see what purpose each is serving logically. Administrative convenience may dictate the way data is held physically; that no longer concerns us.

By cross-referring to the LDS, we should be able to see which data stores serve which entities or groups of entities. For example, on the LDS for Super Systems plc

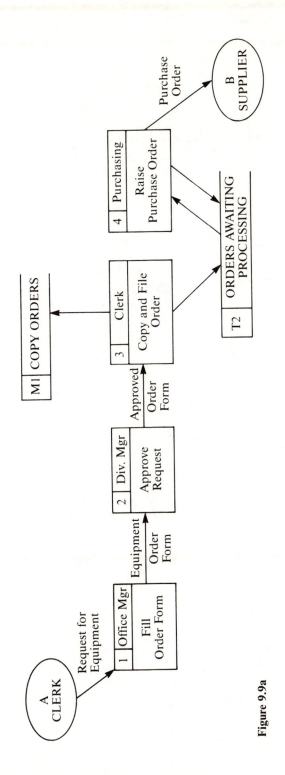

Figure 9.9a

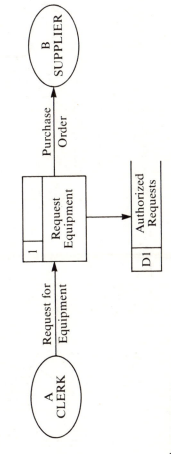

Figure 9.9b

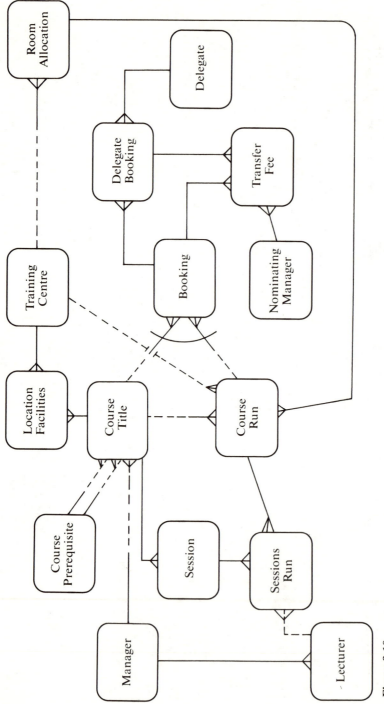

Figure 9.10

we find that a Course Run is related to Training Centre, to Rooms, to Session Run, to Bookings, to Tutors and to Course Title. Our task is to see which of those entities will be placed in the same data store on the DFD as Course Run details. One way to do this is to examine the way in which entries on the data stores are keyed.

If two data stores on the diagram share a common key, consider whether they are in fact the same data store. If so, combine them to form one store, as in Fig. 9.11 below. Here, the data store Wallchart is the only one that is keyed on Course Title Code/Date Commencing. What is held on the Wallchart, though, is information about the location (Training Centre) and Rooms booked for that course. That will help us make a correspondence between Wallchart on the DFD and the entities Course Run, Room Allocation, Training Centre.

The other possible entities: Booking, Course Title, Session Run and Tutor must be similarly examined to see if they belong to the Wallchart information, or other data store subjects.

The data stores Charge Book and Portfolio share the key Course Code Number. It is a matter of administrative convenience that they are kept separately, so in our Logical DFD we combine them to form the one data store Course Type. This has an exact correspondence with entity Course Title on the LDS, and so both physical data stores are replaced on the DFD with Course Title, regardless of physical implementation.

We now have two logical data stores, corresponding with two groups of entities. Continuing the analysis gives us two further logical data stores: Booking (combining Bookings Folder and Wait List) and Tutor, matching on to the physical Tutor Schedule.

Logical data stores are identified only by a reference number, with the prefix D or T. There is no identification of manual data stores at this point.

Redrawing our Level 1 DFD with the data stores rationalized thus, gives us the model in Fig. 9.11.

When we have completed the rationalization of data stores we complete the document: Logical Data Store/Entity Cross Reference. This is a diagrammatic description of which entities are related to which data stores on the Logical Data Model. Figure 9.12 shows the Logical Data Store/Entity Cross Reference for SS plc.

TRANSIENT DATA STORE

A transient (temporary) data store may be a logical necessity, or may only exist for reasons of convenience or policy. If it is the latter, as in data store T1, in Fig. 9.13, it has no place on the Logical DFD and can be removed.

To determine whether it is a logical necessity or not, you must ask whether the task can be performed without it, or would it just be performed differently. If it could not be performed at all, the store is logically necessary and must stay.

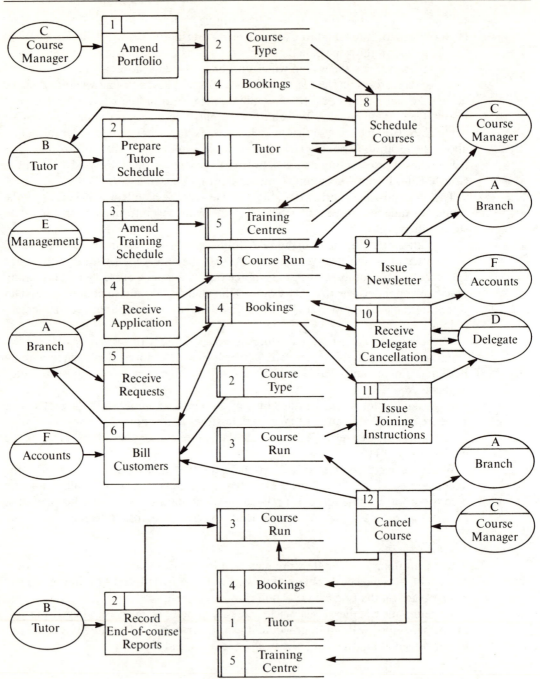

Figure 9.11

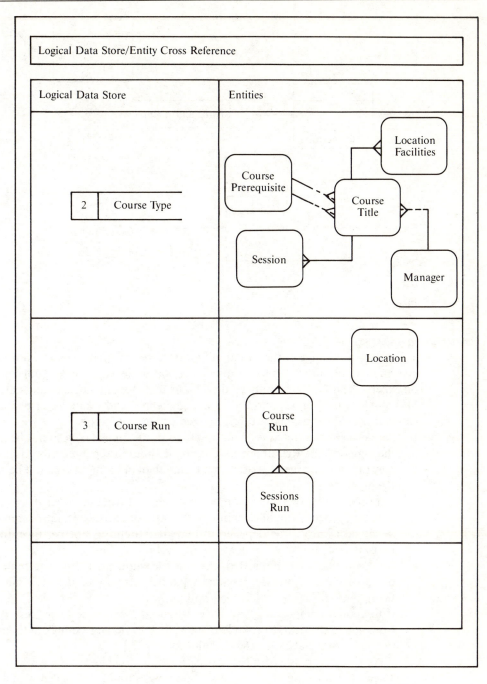

Figure 9.12

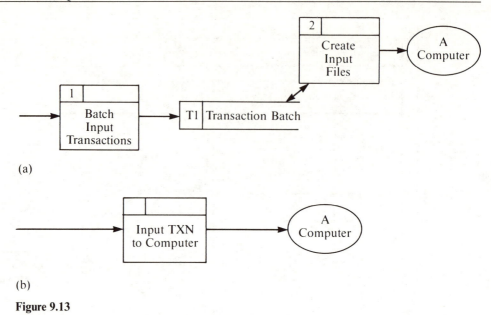

(a)

(b)

Figure 9.13

Rationalizing processes

The nature and range of the situations confronting the analyst make it impossible to give an infallible algorithm for rationalizing bottom-level processes; judgement based on experience is really the only guide. That is no help, of course, to an analyst on his first SSADM project, so below are some guidelines and examples of processes that can and should be rationalized, i.e. merged or removed altogether.

1. Any process that retrieves data purely to print or display need not be shown on the diagram; the Requirements Catalogue should keep a record of it.

2. Any process that merely sorts data, rather than updating it need not be shown on the diagram.

3. If a process involves a decision being taken by a person, rather than automatically, that process should be split into two or more processes; the person involved in the decision will be depicted as an external entity. For instance, in a warehousing system Fig. 9.14a becomes Fig. 9.14b.

4. If two processes are duplicated, examine the whole picture to determine whether or not they need to be represented separately. It may be that you can combine them, but only the complete view will tell you.

5. If in your decomposition you have produced a sequence of processes, moving from one to the next, to complete one task, these can be combined to just one or two processes that define the overall task.

As we have removed all reference to the physical environment, so we must amend the labelling to reflect this. Data flows should be labelled with a succinct description of the flow, not with a document name. Remove references to locations on the process boxes: these reflect the physical, not logical system.

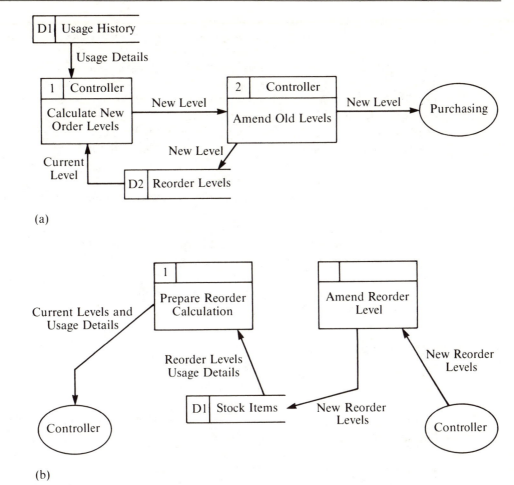

(a)

(b)

Figure 9.14

These examples are guidelines to get you started. You may well find that after you have rationalized data stores and processes your diagrams have about three or four fewer boxes than before. As long as all of the symbols reflect the logic of the system and not the physical, or contingent, constraints, that is all right.

Every stage of DFD production is iterative, so this part will also probably require several passes before it is completed to your satisfaction.

Grouping lower level processes

Processes and data stores are all going to be linked by data flows. Our objective is to link processes in such a way that data flows are kept to a minimum.

Two principles to guide us are:

1. *Maximum cohesion* Keeping together processes that have most data in common.

2.   *Minimum coupling*   Reducing the flow of data to a minimum.

To keep to these principles, identify those processes that access the same data store. A process/Data store matrix will help make this identification.

Having done this, we find that each data store will have a group of processes that access it, some for read-only access, some for write-only access and some for read and write. This will be your first grouping. Now look at all the processes in each group, and further subdivide them into the following:

1. those that are part of the everyday running of the business;
2. those that are periodic, i.e. generated quarterly, or at the end of the financial or tax year, etc;
3. those that maintain reference information.

These groupings will be used again, in the Function Catalogue for the Required DFD, (see Sec. 9.7).

The logicalization process is not algorithmic, but rather heuristic, or rule-of-thumb based. It is not expected that this product would be presented to the Users for validation: after all, they have already approved the Physical DFM. To ask them to approve a stripped-down version of the same system as an accurate picture may be asking too much of their understanding. The Logical DFM is a necessary working document on which you will base the Required DFM.

9.7   Producing the Required DFD

The DFDs of the required system are drawn up first in the creation of Business System Options (BSO) in Step 210. At this point all the user needs is a top-level DFD, or rather, one DFD for each option on the menu.

The Required DFD will be based on the current Logical DFD. To turn this into a BSO, consider, with the aid of the Requirements Catalogue, what processes can be automated, which may/must be manual, what data stores will be required, and any new or altered data flows.

It may be that there will be no drastic differences between the options, but there must be sufficient to show the User clearly what the choices are. As shown in Chapter 17 the BSO will comprise top-level DFDs and rough Cost/Benefit Analyses for each.

Once the user has chosen—and this selection will probably be a combination of features from two or more BSOs—then the top-level DFD is expanded as with the current physical in Step 130. The Logical DFD will be used for reference where possible, to ensure that facilities currently offered are not omitted from the new system, unless through deliberate choice.

When the decomposition is complete, a check should be carried out against the Requirements Catalogue to ensure that all the items on the list have been met by the new processes. Although the DFM for the current system must be rigorous and accurate, the emphasis should be on development of the required model: SSADM is more concerned with meeting the requirements for the future, than in examining the current.

When the Required DFD has been taken to its lowest level, the accompanying documentation must be completed:

- *Data Store/Entity Cross Reference*   See Sec. 9.6.
- *Elementary Process Description*   A brief prose description of each process, to amplify the labelling. (Fig. 9.15)
- *I/O Descriptions*   Every screen, form or print used in the system is to be entered, with data items, format and size. Details of enquiry screens and reports, as well as updates, will be entered. (Fig. 9.16)
- *External Entity Description*   A brief description of each External Entity. (Fig. 9.17)

When this is completed, review, as always, with the User, to ensure that nothing is omitted, and that nothing is unclear.

SUMMARY

Using DFDs and supporting documentation, we have analysed the workings of the current system, reduced it to its *logical* functions and described a new system to meet the stated problems and requirements. This sequence can be represented by Fig. 9.18.

All we have done, however, using this very useful graphic tool, is to *describe*, not specify, the system. We shall use the Required DFDs and documentation later in the design process, when we prepare the enquiry processes and update accesses, and when we carry out Relational Data Analysis. It should be remembered that SSADM uses DFDs primarily as a tool for analysis and communication, not for design.

The design process makes use of the view of the required system described in the DFD of the Business System Option, but other, more powerful tools are used to design the data and procedures.

EXERCISE

Below is an environment description of an information system. Draw a top-level DFD of the system.

Environment

Old Krate's is a travel agency, specializing in exotic holidays. They hold lists of hotels and charter flights, and create bespoke holidays for clients. Bookings are made either through a list of O.K.'s agents, or by direct approach from clients.

Procedures

When a client/agent makes an approach, the reservations clerk selects appropriate flight details and hotel details for the customer and makes a provisional booking. The details are entered onto a Provisional Booking file.

The customer must confirm this booking within three days, by sending a deposit of

Elementary Process Description

Variant: **Current Physical**

Process ID/Common processing ref: **4.3**

Process name: **Update Wallchart**

Common processing cross-reference: **N/A**

Description

The Bookings Clerk receives no. of places booked for a given Course Run, and records on the wallchart the numbers of provisional/confirmed numbers.

After making the update to the wallchart, the Bookings Clerk posts all the delegate details to the Bookings Folder.

Figure 9.15

I/O Descriptions

Variant: *Required System*

Project/System	Author	Date	Version	Status	Page of
SS plc					

From	To	Data flow name	Data content	Comments
D	7.1	Notice of Cancellation	Course Code Course Date Delegate No. Branch No. Manager ID	Input to process
7.1	7.2	Cancellation Notice	Course Code Course Date Delegate No. Branch No.	
7.2	MS	Old Delegate Details	Course Code Course Date Delegate No.	
7.1	7.3	Request for Standby	Course Code Course Date Branch No. No. of places	
7.3	D	Offer	Course Code Course Date New Delegate No. New Delegate Name New Delegate Address	
M3	7.3	Candidate Details	Branch No. Branch Address Delegate No. Delegate Name	

Figure 9.16

External Entity Description

Variant: **Current Physical**

Project/System SS plc	Author	Date	Version	Status	Page of

ID	Name	Description
A	Branch	A regional unit of SS plc, that places requests, via nominating managers, for training for its members.
B	Tutor	An employee of SS (Training) plc, who delivers sessions on Scheduled Courses.
C	Course Manager	The owner of a Course Title, who is responsible for providing details of courses, and making the decision to cancel Scheduled Courses.

Figure 9.17

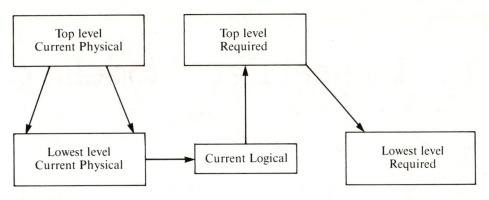

Figure 9.18

10 per cent of costs. On receipt of this deposit, Reservations close the Provisional Booking and add the details to their Full Booking file.

Four weeks before the flight is due Accounts send an invoice to the client for the balance. Accounts notify Customer Services when the balance is received, and Customer Services then send Tickets and joining instructions.

Reminders are sent to customers three weeks and one week before departure. Although the company insists that payment is made at least one week before departure, it has been known that payment has been made and tickets received on the morning of a flight.

At the end of each month, commission of 15 per cent is paid to any agents responsible for holidays commencing during that month.

10. Logical Data Modelling

10.1 Aims of chapter

In this chapter you will learn:
- The use of Logical Data Modelling in SSADM.
- How to identify entities.
- How to chart and label relationships.
- How to show optionality in relationships.
- How to cross-validate the DFD with the LDS.
- How to validate Logical Data Structures with Enquiry Access Paths.
- How to document Logical Data Structures.

10.2 Where Logical Data Modelling is used in SSADM

The LDM is created early in the SSADM cycle, and updated at several points subsequently.

Step 110 Produce overview LDS.

Step 140 Produce an outline LDS of the data in the current system and identify attributes.

Step 150 Convert DFD of current services to logical view. Use LDM to define logical data stores.

Step 210 Prepare BSOs and support with LDM.

Step 310 Use LDM to update low-level DFDs with new data items.

Step 320 Extend current model to support new processing requirements. Fully document entities and attributes.

Step 340 Carry out Relational Data Analysis on I/O Descriptions, and use the results to validate and enhance the LDM.

Step 360 Use the LDS to create the Entity Life Histories in two passes—bottom-up first and then top-down for showing abnormalities. The LDS may be extended or amended in the light of this activity.

Step 370 Review the Requirements Catalogue for both functional and non-functional requirements. Amend the required LDM as appropriate.

Step 520 Complete the process definitions for each update function, using the ELHs, Required System LDM and Effect Correspondence Diagrams (see Chapter 13).

Step 530 Process Definition for all enquiries is completed here, with enquiry aspects of Update Processing. Inputs are the Required System LDM and the Enquiry Access Paths.

Step 620 The Required System LDM is converted into a format which is implemented on a product-specific DBMS design.

Use in SSADM

Inputs to LDM are as follows:
- Feasibility Report
- Project Initiation Document
- Overview LDS
- Requirements Catalogue
- Selected Business System Option

Outputs from LDM are as follows:
- Logical Data Structure
- Attribute Description
- Entity Description
- Domain Description
- Relationship Description

Figure 10.1 illustrates the relationship between LDM and other SSADM techniques.

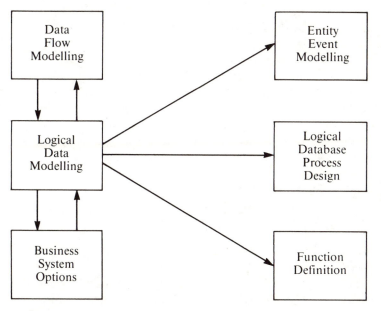

Figure 10.1

10.3   Logical Data Structures

Data Flow Diagrams of the current system show us *how* data is stored in the workings of the system, but this is usually in a very unstructured manner. By 'unstructured' I mean that ad hoc storage of data items, especially in a manual office, put data items together for convenience rather than by logic.

Let us take details of one of SS plc's courses for example. If we look at the form that is used to make bookings (Fig. 10.2), we see that we have information on Courses, Scheduled Courses, Locations, Managers and Delegates, all on the same piece of paper.

If we were to store this Booking Request on computer as it stands, we would have a very strange record, because it would not be possible to say what it is a record of; Manager? Delegate? Course? etc.

Logical Data Modelling is a technique for examining the *structure* of the data that is hidden inside such documents. It is *logical* because it ignores the physical implementation of the storage in our system.

The technique is, like most of SSADM, based on a diagrammatic notation, for ease of creation, amendment and understanding. The notation consists of just two components, representing *entities* and *relationships*.

Entities

By 'entity' we mean something in our system about which we store and use information. On our Booking Request form above, we asked, what is the subject, Manager? Delegate? Course? The fact that we can ask such a question points out each of those as entities.

An entity must have the following characteristics:

- It must contain information of interest to the system.
- There must be the possibility of more than one occurrence—the environment itself (SS plc) cannot be an entity as there is only one of it.
- Each occurrence must be uniquely identifiable—there must be a code or key for each entity.

While an entity often refers to something tangible, such as a person, it may describe all sorts of other kinds of subject. Don't look only for the obvious physical objects. Likely kinds of entities may be:

- Tangible, e.g. Person, Product.
- Conceptual, e.g. Course Title.
- Active, e.g. Delivery, Course Run.
- Permanent, e.g. Warehouse.
- Volatile, e.g. Stock.

IDENTIFICATION

It has already been stated that each occurrence of an entity must be uniquely identifiable. This implies the use of a code. When carrying out your investigation of the environment, look for any code numbers in use. If the code can identify such an occurrence uniquely, you have found an entity.

Booking Form

Super Systems : Training

SST Course Booking Duty:	

Course Title

Delegate Details (Please advise of special requirements/disability):

Surname

First Name

Division ☐   Section ☐   Group ☐

Nominating Manager:

Duty

Telephone

Branch

Signature	Date

Figure 10.2

Common codes in use are such things as Personnel Number, Part Number, Account Number, Invoice Number, etc. Many codes will in fact be combinations of other codes. If this is the case, check to see if each part is unique, or common.

Take the example of a company handling air freight. Let us postulate a document, Airbill, which identifies a consignment for a particular airline, stored in a particular shed. The code which identifies each one is made up of: Airline code, Shed Code, Bill Code. If Bill Code is unique within the system, then we have found *four* entities, Airbill, Airline, Shed and Bill.

If on the other hand Bill Code is repeated for each airline, then we have still found four entities: Airbill, Airline/Bill, Airline and Shed. In this case Airline/Bill has a *composite* key, i.e. one made up of two or more items, one of which is not unique in the system, and at least one of which is unique in the system.

When identifying entities by keys, therefore, look for the simplest identifier for a group of data items. That may be one data item alone, as in Airline, or it may be a composite, as in Airline/Bill. In both cases, the identifier is the simplest possible.

Relationships

A relationship is a necessary association between two, and only two, entities. For example, let us identify two entities from SS plc:

1. *Course Title*, with a key of the first three letters of the course title (Course Code), is an entity describing the titles in the brochure.
2. *Course Run*, with a key of Course Code and Start Date, is each occasion that that course type is presented to a class of students.

There is a necessary relationship between these two entities in that Course Run refers directly to a specific Course Title, and Course Title is given several presentations, or runs, to classes. Each relationship has several features that must be considered. They are:

- cardinality, or degree,
- optionality and
- names.

CARDINALITY (DEGREE)

This term simply means one of the following facts about a pair of entities: one occurrence of an entity in the relationship is associated with just one occurrence of the other (1:1), or each occurrence of one entity in the relationship can be associated with more than one occurrence of the other (1:m). The entity at the 1 end of the relationship is termed the *master* and the entity at the m end is termed the *detail*. Finally, occurrences of either entity in the relationship may be associated with more than one occurrence of the other (m:n).

That end of a relationship line that is attached to a detail entity, i.e. one with many occurrences, has a 'crow's foot' attached, to denote 'many'. Figure 10.3 tells us that one Course Title will have many Runs, while each Run of a course will be for one and only one Course Title.

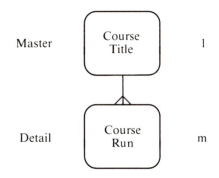

Figure 10.3

Permitted cardinality
In the design modules of SSADM, only the 1:m structure is accepted as being valid. If, therefore, investigation suggests the existence of 1:1 or m:n relationships, these must be analysed further and resolved as explained below before carrying out Physical Design.

1:1 relationships   In many cases, 1:1 relationships are in fact a description of just one entity: the other entity merely contains extra information rather than information about a different topic. When this is the case, the entities should be merged to form one.

If this is not the case, the relationship should be shown as a 1:m instead. One of the entities in the relationship will have been formed first, so make that the single occurrence, and the one created second becomes the many end with the crow's foot.

An example of this would be a situation where a company invoiced customers for goods delivered rather than goods ordered, in other words, presented the bill on delivery, even if it is only a part delivery. In this case, there is a 1:1 relationship between Delivery and Invoice. As the Delivery entity would be the first created, that

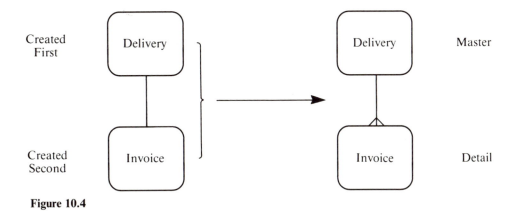

Figure 10.4

would become the 1 end (or master), and the Invoice would become the many end (or detail). Figure 10.4 shows this.

m:n *relationships*   If you uncover an *m:n* relationship, the odds are that there is another entity that you have not yet recognized. In SS plc, we find that a branch makes a number of bookings for a number of our courses. Each Booking may be for a number of Delegates, but for one Course only. Each Delegate may be booked on a number of courses, i.e. be the subject of a number of Bookings.

At first glance, we identify the following entities: Course, Booking, Branch and Delegate. This is modelled in Fig. 10.5.

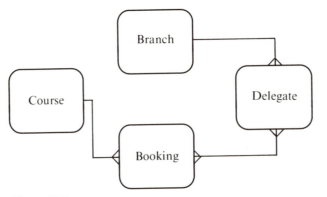

Figure 10.5

We have an *m:n* relationship between Delegates and Bookings, which must be resolved. There is information on each Delegate, such as Name, Contact Address, Special Requirements and Telephone Number. There is information on Bookings, such as Authorizing Manager, Date of Booking, Contact Name and Number of Delegates. What we do *not* have, though, is a way of identifying a particular booking for a given Delegate.

This tells us about the missing entity that will resolve the *m:n* relationship. The entity acts as a link (and so is called a *link entity*) between Booking and Delegate, and for its key will take a combination of Booking No. and Delegate No. Naming the entity sometimes causes a problem, but in this case Delegate Booking will clearly indicate the nature of the entity.

When a link entity is identified and inserted into the diagram, it becomes the detail of the two entities that is reconciled, as shown in Fig. 10.6.

Sometimes the link entity will contain additional information in the form of data items (called *attributes*) that are not present on either of the two original entities, and sometimes the link entity itself is all the extra information held.

In our current example, Delegate Booking tells us nothing new about the Delegate or the Booking *per se*. Its extra information value lies in identifying the one Booking out of many for a particular Delegate.

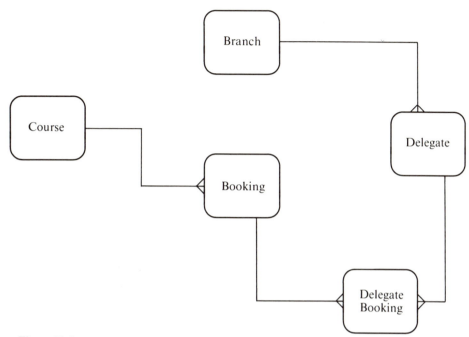

Figure 10.6

On the other hand, if we look at an Order Form in a Mail Order Processing system, we can identify an $m:n$ relationship between Order and Catalogue Item, in that each Order form will be for many items, and the Item will be the subject of many Order Forms, as shown in Fig. 10.7.

Figure 10.7

The link entity in this case will be Order Line, with a key of Order No. and Line No., or perhaps Order No. and Item No. Whichever of these keys is chosen, the link does contain an important item of information that is not on either of the other entities, and that is the Quantity of that Ordered Item on that one Order. Figure 10.8 shows this resolution.

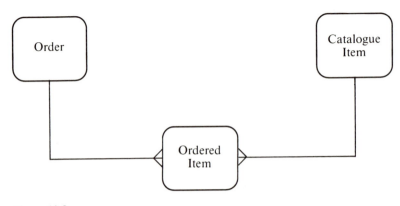

Figure 10.8

OPTIONALITY

A relationship between entities may always exist as soon as one of the entity occurrences comes into existence. Figure 10.9 shows a Booking for many Delegates for a Course Run. As soon as a Booking is created, there is a necessary link between Booking and Course, and Booking and Delegate. It cannot exist without at least one of each being attached to it. This means that the relationship between Booking and the other two is *mandatory*.

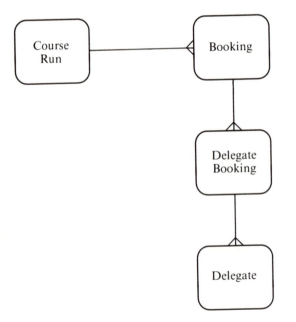

Figure 10.9

Other entity occurrences in a relationship may exist even if there are no occurrences of the entity in existence at the other end.

Using the same example, a Course can be scheduled to run, and so the occurrence of Course Run exists before any bookings have been made. In this case, the relationship between Course Run and Booking is said to be *optional*.

(= **May** have)   (= **Must** be for)

Figure 10.10

A mandatory relationship is depicted by a solid line leaving the entity. An optional relationship is depicted by a broken line leaving the entity. In Fig. 10.10 the relationships are read as follows:

● A Booking *must* be for one and only one Course Run.
● A Course Run *may* have one or more than one Booking made for it.

There is another instance of Optionality that should be highlighted, and that is the possibility of mutually exclusive relationships. By that, I mean an instance where a master entity has two possible detail entity types, but each occurrence of the master will have only one of those details associated with it.

In SS plc, a Delegate may be booked on a course as part of his/her training programme. The booking will be either for a Course Title, or for a Course Run. If it is for the Course Title, the delegate will be on a Waiting List; if it is a Course Run, it will be a Confirmed Booking. It cannot, however, be both.

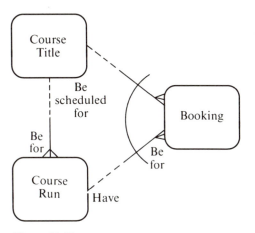

Figure 10.11

We represent that on the model by crossing the possible relationships with an arc, as in Fig. 10.11. That identifies an exclusive relationship, either from one master to two possible details, or from one detail to two possible masters.

NAMES

One way of introducing rigour into LDM is by forcing the analyst to name each end of the relationships. It is not very useful to say simply, 'Each occurrence of Entity A has one Entity B attached to it.'

To show that we have fully understood the nature of the entities and their relationship, both master and detail ends should be described in a succinct manner. Find a meaningful word or phrase that connects them, avoiding the weak 'has', and use the following formula:

- Each occurrence of Entity A must/may (link word) with (or 'to', or 'for', etc.) one or more occurrences of Entity B.

for one end, and for the other end:

- Each occurrence of Entity B must/may (link word) with/to/for one and only one occurrence of Entity A.

The selections 'must' or 'may' in each of these statements reflects whether that relationship end is optional or mandatory. If the relationship is mandatory use the word 'must', and if optional, use the word 'may'.

Figure 10.12 shows the relationship between Course Title and Course Run.

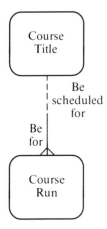

Figure 10.12

The relationships are named as follows:

- Each Course Title *may* be scheduled for one or more Course Run.
- Each Course Run *must* be for one and only one Course Title.

Domain

A domain is a permitted range of values for an attribute. Date of Birth, for example will have a domain that is different for particular entities; a baby in a clinic will have a domain that goes back no further than say three years. A student at a college, on the other hand, will have a domain for Date of Birth that will be no more recent than the previous sixteen years.

A serial code of five numbers will have a domain of any possible combination of five numeric characters, or to put it more formally, 00,001–99,999. If the code in the system only covers the range 01,000–50,000 for some reason, then that range is the domain for that code.

10.4   How to derive the LDM

While there is no rigid algorithm for creating a Logical Data Structure, the following guidelines will help the beginner, as well as providing experienced practitioners with a checklist of activities.

Identify the candidate entities

During the fact-finding exercise at the start of analysis in Step 110, identify as many codes as you can that are used in the system, or will be used if this is a greenfield system.

Having listed the candidate entities, criticize the list with the users. Be careful to identify synonyms (two names for the same logical entity, perhaps by separate departments) and homonyms (one name for two separate logical entities).

Once you have identified entities, be sure that the users are happy with the names chosen. The entity will not reflect the way that the data is held in the physical system. For example, an order processing system may hold data stores for Received Orders, Pending Orders, Partially-filled Orders and Filled Orders. These data stores may be scattered across two or three rooms. All, however, refer to the business entity Order. Do not create separate entities for Pending Order and Filled Order: identify the underlying entity Order and let that summarize all the different data stores relating to it.

Further entities will be identified later, during Step 340, for instance, using RDA, and during Step 360 using Entity Life History analysis.

Remember the feature that an entity must have the potential for more than one occurrence: lists and catalogues are not themselves entities; each entry may be.

Identify relationships

Very small systems (those with a dozen or so entities) pose little problem when identifying relationships. Large systems, though, can include several hundred entities, with necessarily complex webs of relationships.

One way of identifying relationships between entities is the use of the matrix. The larger the system, the more advisable this is. It enforces a rigorous examination of

	Course Title	Training Centre	Delegate	Course Run	Booking	Nominating Manager	Tutor	Room Type
Course Title	X			X	X		X	X
Training Centre				X				
Delegate					X	X		
Course Run					X		X	
Booking						X		
Nominating Manager								
Tutor								
Room Type								

Figure 10.13

every pair of entities for a necessary and direct association. Figure 10.13 gives an example of a grid, based on a subset of SS plc.

The marking of the grid is a matter of local standards. A relationship can be noted by a simple *y* or *x* on the intersection; it can be shown more rigorously by indicating the cardinality on the intersection, with an *m:d*, stating which was master and which was detail.

Having identified the relationships, the more difficult tasks of naming each end, and agreeing optionality, remain. This must be done, as always, with the help of the user.

Draw the Logical Data Structure

The first LDS is drawn in Step 110, but this is a crude overview. Step 140 refines and completes this picture as the fact-finding exercise progresses. The LDS for the required system develops during Stage 3, particularly in Steps 320, 340 and 360.

There are no hard-and-fast rules for drawing the structure, except that you must be prepared to make several attempts before you and the users are satisfied. There are a few guidelines, however, to simplify the process:

● Those entities that have the most relationships should be placed at the centre of the page.

- Those entities that have no masters (they are known as reference entities) should be at the top of the page.
- Let natural hierarchies of relationships cascade down the page.
- Try to avoid crossing lines: logically there is nothing wrong with them, cosmetically they can make the model hard to read.
- Name all relationship ends on the model.

Normalize the LDS

Full normalization (see Chapter 12) does not take place until Step 340. However, to help identify or clarify some entities, it can be done informally in Step 140. The normalization process is not described here.

Validate the LDM against functional requirements

The LDM must be consistent with the DFD on two counts: processes and data stores.

- *Processes*   Every entity on the LDM must be capable of being created, deleted and amended. The DFD should show processes that do all of these for each entity.
- *Data stores*   Ensure that after carrying out logicalization of data stores that each entity is represented in only one Logical Data Store.

Once the DFD and LDM have been cross-checked, and one or both amended accordingly, validate the access paths around the model, to support the requirements. For this, use the Elementary Process Description (EPD), and ensure that the model can provide all the data items to perform each process.

For every EPD on the Data Flow Model there must be an access path on the LDM. This will consist of an entry point into the diagram, such as a key, and a valid navigational path around the relationships. The aim is to ensure that every data item named on the EPD can be accessed—whether to create, update, delete or read—on the LDS. The data is accessed through one or more of the following:

- Read entity directly (using given key).
- Read next detail from found master.
- Read master from found detail.

If it is not possible to access all the items, either the EPD or the LDS is wrong. Consultation with the User is necessary to clarify this.

This activity is carried out informally in Steps 140 and 150 for the current model, and Step 320 for the required model. A more formal access path analysis is carried out in Step 360.

Remove redundant relationships

After validating the model against requirements, we must identify redundant access paths (relationships). Look for any 'enclosed' structures on the model. Figure 10.14 is an example of just such a structure, and illustrates the relationships between Course Title, Course Run, Delegate Booking, Delegate and Booking.

We have three enclosed structures in this subset; all need to be examined. Let us

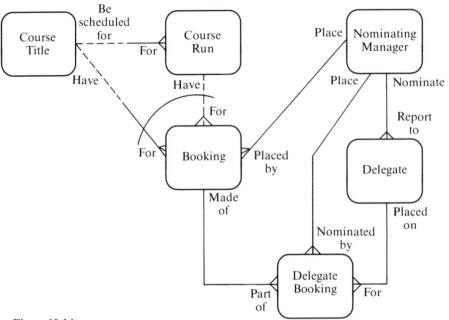

Figure 10.14

take the requirement to Send Joining Instructions to Nominating Managers of all Delegates for a forthcoming Course Run.

Access is on Course Run, retrieving all those whose Start Date is the same as the target date. For each of those Course Runs, retrieve each Booking.

The last access can be achieved in three ways: via Delegate Booking to Nominating Manager, or Delegate Booking to Delegate to Nominating Manager, or straight from Booking to Nominating Manager. Information is needed on Delegate Booking, Delegate and Nominating Manager to send out the Joining Instructions, so in any event we must find all three entities.

In each access, the same information is available, so is there any advantage in having three access paths? The answer is no, and so we delete one path. In this case the link between Nominating Manager and Delegate Booking serves no obviously useful purpose, and so we can dispense with it. In removing it we lose no information. The same information is obtained from either of the other paths.

Similarly, we can ask if either of the other two access paths is redundant. We can see that by going from Booking to Delegate Booking to Delegate to Nominating Manager we can obtain all the information we need. Going from Booking to Nominating manager and down, on the other hand, does not identify which delegates require Joining Instructions for the given course. All we can identify there is that the Nominating Managers have some Delegates reporting to them, and these Delegates are booked on some courses. The relationship between Booking and Nominating Manager, therefore, is the more dispensable of the two. So it goes.

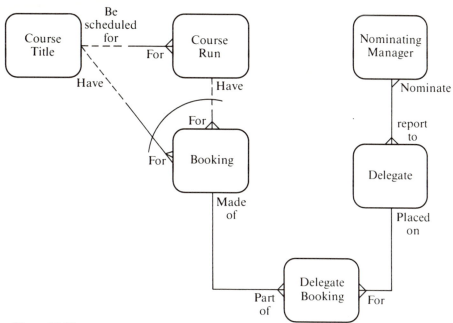

Figure 10.15

After this process, our structure is simpler, and looks like Fig. 10.15.

One word of warning: not all enclosed structures contain redundant relationships; all are candidates for such examination, but some need all the relationships to fulfil the requirements.

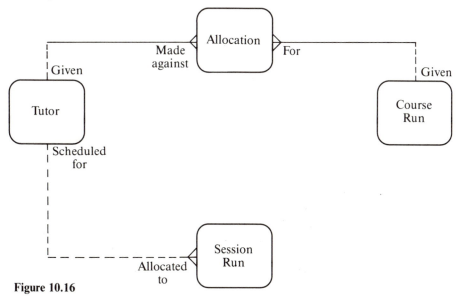

Figure 10.16

Beware of what is known as the 'connection trap', or, more simply, as the 'body of the crow'. Figure 10.16 illustrates this. If we want to see which tutor was used for a given session on a particular Course Presentation, we cannot find that information from this structure. We can come in on the Course Run and find from that the tutor concerned. We can then find every Session presented by that tutor, but there is no way of linking any one of those presentations with a particular Course Run. To achieve that, we must put in a relationship between Course Run and Session Run.

Be wary, then, of rationalizing every enclosed structure—in this instance the structure is necessary. The rule is to examine all the requirements to make sure that you are not losing information by removing a relationship.

Define Enquiry Access Paths

We have already validated the model against the functional requirements; in Step 360 we validate the Required System LDM against the Enquiry Function Definition. This technique of validation is more precise than the earlier validation, and formally defines a set of documented Enquiry Access Paths, which will be input to Step 530, Produce Enquiry Processing Models.

Enquiry Access Paths have their own diagrammatic notation, loosely based on Jackson techniques, that outline the access to given entities to satisfy requirements.

The analysis is carried out on the Required System LDM. The procedure to follow is:

1. Draw the part of the LDS needed for the enquiry, either because it contains the information, or because it is needed as a navigation point across the structure to the data needed.
2. For each enquiry that accesses entities in a relationship:
    (a) Show each access from master to detail vertically.
    (b) Show each access from detail to master horizontally.
3. What we have now is a subset of the LDS, with horizontal and vertical access paths. Next, we need to convert this view into an Enquiry Access path.

    To do this, remove the crow's foot relationships, and connect entities with an arrow instead, called an *access correspondence arrow*. This arrow shows an access path between two different types. It can indicate the sequence of accesses carried out, or it can access from one entity type to a set of details; this set is denoted as an iteration.

    If an enquiry wants to regard occurrences of one entity type differently, depending on an attribute, show this as a selection. Thus, a Course booking may be a Provisional Booking or a Confirmed Booking. If an enquiry needs to distinguish them, Course Booking will have underneath it a selection of Provisional or Confirmed.

    If the enquiry requires that an occurrence of an entity, or piece of information, is presented in a particular order, (e.g. date order), and that order is not met by the access sequence, make a note on the EAP that a sort process is also required.

    Mark the entry point for the enquiry with the criteria, e.g. the key, or search criteria.

4. Confirm that the required data can be obtained in one of the following three
    ways:
    (a) Read record directly using entry attributes.
    (b) Read next detail from current master.
    (c) Read master from current detail.

Let us take an example from SS plc: generate the timetable for a Course Run.
Sessions will be generated from standing data about the course. Tutors for each
session will be named on the timetable; any qualifications they hold (MA, Ph.D,
MBIM, etc.) will be retrieved from their file and entered by their name on the
timetable. Location address and room numbers will also be entered as header
information on the timetable. Figure 10.17 shows us the required view for this
requirement. Figure 10.18 shows us the master and detail relationships converted to
horizontal and vertical access paths.

Now we redraw this section of the model as a Jackson structure, grouping the
correspondences at the same time, giving Fig. 10.19. We can confirm that all of the
enquiry elements are met by direct read, master-to-detail access or detail-to-master
access.

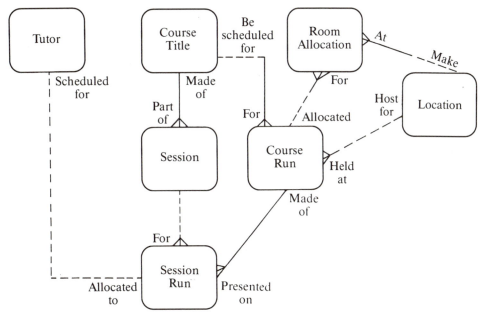

Figure 10.17

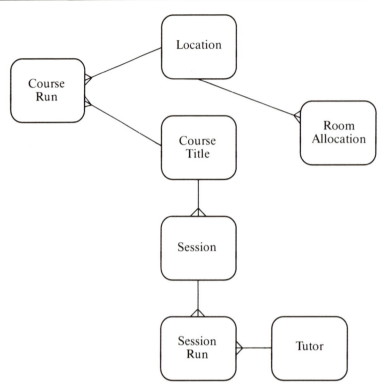

Figure 10.18

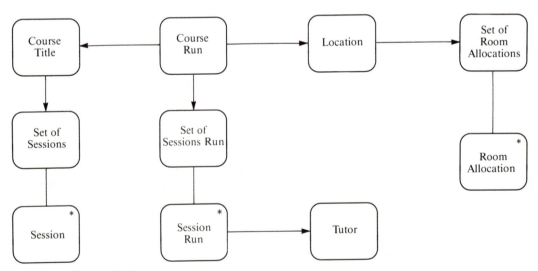

Figure 10.19

Presenting the LDM to the User

Quality assurance reviews are not built into SSADM as such, although they are inherent in the Project Management method that drives the SSADM project. However, it is important that the User accepts the LDM as accurate and valid before the formal QA procedure. We will, therefore, show it to the User informally for comment and confirmation. As most SSADM systems are very large, only show the Users a subset at a time, rather than the whole model: remember, it is a tool for communicating with the Users as well as an aid to our own understanding, so do not give them something too complex and superfluous for the immediate purpose.

Although at each presentation you will show a subset of the model, the entities shown will have other relationships that are irrelevant to the presentation. Show the connected entities as broken boxes, as in Fig. 10.20, just to prove that although outside the scope of the subset, they are still there in the large model.

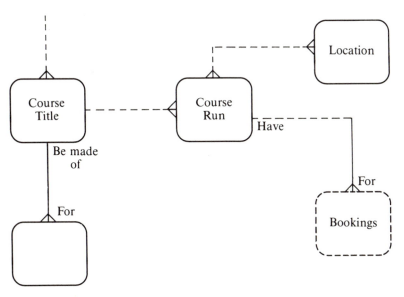

Figure 10.20

Document the LDM

So far only the Logical Data Structure diagram has been discussed. The full Logical Data Model comprises the LDS, Entity Descriptions, Relationship Descriptions, Attribute Descriptions and Domain Descriptions. Relationship Descriptions will comprise two sheets, one for each end of the relationship.

As the model develops in detail, these descriptions too must be completed. The descriptions are on a standard SSADM form, and incorporate both functional and operational descriptions.

Entity Description — Part 1

Variant: Current

Project/System SS plc	Author	Date	Version	Status	Page of

Entity name Course Run	Entity ID

Location N/A	Occurrences 1000	Average 2500	Max.

Description   The actual scheduled presentation of a Course Title

Synonym(s)   Scheduled Course

Attribute name I/D	Primary key	Foreign key
Course Title	✓	
Course Date	✓	
End Date		
Location		✓
Max Places		
Provisional Bookings		
Confirmed Bookings		

Ref no.	'must be'/ 'may be'	'either'/ 'or'	Link phrase	'one and only one'/ 'one or more'	Object entity name
1	must		Composed of	one or more	Session Run
2	must		Presentation of	one & only one	Course Title
3	may		Allocated to	one & only one	Location
4	may		Subject of	one or more	Booking

Notes

Figure 10.21a

Entity Description — Part 2

Variant: **Current**

Project/System **SS plc**	Author	Date	Version	Status	Page of

Entity name **Course Run**	Entity ID

User role	Access rights
Course Scheduling Bookings	I, D R, M
Owner **Course Manager**	

Growth per period

Additional relationships

None - all shown on LDS.

Archive and destruction

Archive 3 months after course ends.
Destroy 18 months after Course Run ends.

Security measures

No special requirements other than access
limitations above.

State indicator values

1. Course scheduled to run.   4. Provisional Booking cancelled.
2. New Provisional Booking made.   5. Confirmed Booking cancelled.
3. Provisional Booking confirmed.

Notes

Figure 10.21b

Relationship Description

Variant: **Current**

Project/System **SS plc**	Author	Date	Version	Status	Page of

Entity name **Course Run**	Entity ID ╱

Mandatory ☑   Optional ☐   % Optional ☐

Link phase   **must be for**

Description   **Describes occurrences of presentations of Course Title. Describes Start and End Dates, Location and Tutor.**

Synonym(s)   **Scheduled Course**

Object entity name **Course Title**	Object entity ID ╱

One (1): ☑ Many (m): ☐	Minimum	Average	Maximum
Cardinality description ╱			
Growth per period ╱			
Additional properties ╱			

User Role	Access rights
Course Scheduling **Course Maintenance** **Course Bookings**	**Write** **Write** **Read**
Owner **Course Manager**	

Notes

Figure 10.22a

Relationship Description

Variant: **Current**

Project/System **SS plc**	Author	Date	Version	Status	Page of

Entity name **Course Title**	Entity ID ⟋

Mandatory ☐	Optional ☑	% Optional **10**

Link phase   **may be scheduled for**

Description   **The Course Title is described in terms of code, intentions of course, duration, cost, effective date**

Synonym(s)   **Course, course Type**

Object entity name **Course Run**	Object entity ID ⟋

One (1): ☐	Many (m): ☑	Minimum **0**	Average **4**	Maximum **12**

Cardinality description

Growth per period   **In total, approx. 500 per month**

Additional properties

User Role	Access rights
Course Bookings	**Read**
Course Manager	**Write**
Course Scheduling	**Write**
Course Maintenance	**write**

Owner   **Course Manager**

Notes

Figure 10.22b

Attribute/Data Item Description

Project/System	Author	Date	Version	Status	Page of

Attribute/data item name Delegate Duty	Attribute/data item ID Delegate ID

Cross-reference name/ID	Cross-reference type
	Entity Description I/O Structure Dialogue

Synonym(s) Del. Duty   Del. ID

Description

The unique identifier for each Delegate who attends any course. The format is AAA 9999 – Branch Code, Serial Identifier

Validation/derivation

Alpha component must have a match with a Branch Code in current use.

Mandatory ☑ Default value	Optional ☐ Value for null
Logical format Alpha (3) Numeric (4)	Unit measure N/A
Logical length 7	Length description

User Role	Access rights
Bookings A/c	I R

Owner   Accounts Manager

Standard messages

Notes

Figure 10.23

Entity and Relationship Descriptions, and volumetrics, begin in Stage 1. Full descriptions are completed in Step 320, when the required LDS is drawn. In Step 360, the final and complete LDM, with all volumetrics and supporting operational documentation is presented, as well as all functional documentation.

Figures 10.21–10.23 give an Entity Description for Course Run, a Relationship Description for the relationship between Course Run and Booking, and supporting Attribute and Domain Descriptions.

SUMMARY

Logical Data Models provide us with a second view of the system: the underlying structure of the data, and relationships between logical groups of data.

It is developed first of all in parallel with the DFD, in Feasibility, then in Stage 1. On selection of the BSO, the LDM is expanded to support new functional requirements, and is further enhanced by Relational Data Analysis.

The Logical Data Structure itself provides us with a simple diagrammatic notation, but it must be rigorously supported with documentation describing fully entities, relationships, attributes and domains.

EXERCISES

1. The following text describes a college library environment.

    A college library holds books for its members to borrow. Each book may be attributed to one or more authors. Any one author, of course, may have written several books. Up to 10 copies may be held of popular titles.

    A member may borrow up to six books at a time. If books are not returned on time a fine will be levied. If the books are not returned by a week after a third reminder, and fines are not paid, the member may be put on to a blacklist until the situation is remedied.

    If no copies of a wanted book are currently in stock, a member may make a reservation for the title until it is available.

    (a) Draw a Logical Data Structure for this environment.
    (b) Draw Enquiry Access Paths to fulfil the following requirements:
       (i) List all members who have reserved a particular book.
       (ii) List all members on the blacklist.
       (iii) List all the books written or co-written by a specified author.

2. The text below describes the environment that exists within a rail transport authority.

    The authority divides its services into lines (e.g. cosmopolitan line, peripheral line, etc.). Each line is served by a number of depots (a depot only serves one line) at which trains, drivers and guards are based.

    Each line has many stations, and because each line crosses all other lines at some point in the network, a station may be on many lines.

    Lines may have branches at either end, so each line may have many routes. A route is described as running between two stations, in one direction only. Thus,

Route 1 might be from Station A to Station M, while Route 7 might be from Station M to Station A. A route will not embrace more than one line.

Drivers and guards only work on trains based at their depot. Every journey is assigned a driver, but on some journeys, guards are not considered necessary. A journey is a particular occasion of a train travelling a route.

(a) Draw a Logical Data Structure for this environment.

(b) Draw the Enquiry Access Path to meet the following requirement: identify all trains, drivers and guards assigned to a specified line.

11. Function Definition

11.1 Aims of chapter

In this chapter you will learn the following:
- The meaning in SSADM of the term *function*.
- The different ways of classifying functions in SSADM.
- The composition of Function Definition.
- How Function Definition is achieved in SSADM.
- The supporting documentation for Function Definition.
- The Universal Function Model.
- How Function Definition is related to other SSADM techniques.

11.2 Where Function Definition is used in SSADM

Step 330   The different elements for this technique are gathered and identified after the creation of Required System DFDs, in Step 330. We identify the first functions from this model. For each function we identify, we produce I/O Structures, which will be the input to Relational Data Analysis in Step 340. A User Role/Function Matrix is produced to identify which Users are responsible for each function.

Step 350   The User Role/Function Matrix created in Step 330 is used as input to prototyping the I/O interface. The function entries on the matrix are a result of the function identification described above.

Step 360   The Function Definitions described in Step 330 are amended and increased by the activity of Entity/Event Analysis in this step. As new events are identified, the Function Definitions are amended, and the related activities of Steps 330 and 340 are carried out for them.
   In Step 360, we develop Enquiry Access Paths for enquiry functions and those parts of update processes that are enquiry. The principal inputs to this activity are the I/O Structures.

11.3 Functions in SSADM

In SSADM the term *function* applies only to the *new* system, not the current environment. A function is defined as 'A distinct piece of the processing carried out in the new system, as perceived by the User.'

The important part of this definition is that it describes the *User's* view of the system, not the analyst's or programmer's. Functions play a significant part in analysis by drawing together all the products from SSADM that define such a piece of processing.

All of the functions identified in this stage will be carried forward to Stage 5, for detailed Logical Process Design.

Inputs to Function Definition are as follows:
● Logical Data Store/Entity Cross Reference
● Elementary Process Descriptions
● I/O Descriptions
● Required System Data Flow Diagrams
● Requirements Catalogue
● User Roles

Outputs from Function Definition are as follows:
● I/O Structures
● Function Definitions
● Requirements Catalogue
● User Role/Function Matrix

Figure 11.1 illustrates Function Definition in SSADM.

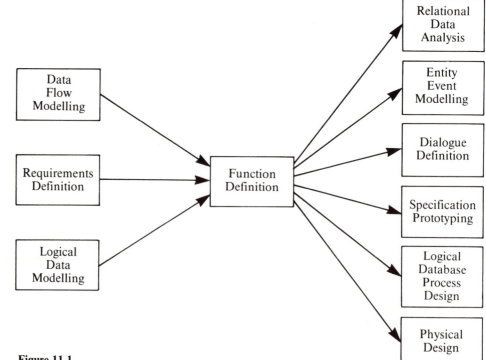

Figure 11.1

Classification of functions

Functions can be recognized as belonging to one of three types:

1. *Process type*   Update or enquiry, i.e. an update to the database takes place, or data is read without change, respectively.

2. *Implementation type*   Online or offline. Many functions are clearly one or the other, but many can be input and processed in either mode. If a function is implemented either way, a separate function for each should be defined, with cross reference between them.

   If a function can be initiated by an online input, but is then processed later, offline, make a choice as to how it is to be classified, one or the other.

3. *Initiation type*   Is the function initiated by a deliberate input from the User, or does the system automatically trigger its execution? A system-triggered function could be because of time, perhaps, say month-end or quarter-end. Alternatively, it could be because of a condition in the data, such as a customer dangerously exceeding a credit account and cautionary action being taken.

11.4   Procedures

Identify functions

The essential inputs to the initial function identification are the DFM of the new system, the associated I/O Descriptions, the LDM of the new system and the Requirements Catalogue. As stated above, the identification takes place in Step 330 in the first instance.

Look at the DFD for the required system. Look for all the original triggers of processing, i.e. flows into the system from an external entity, or a process triggered by a data flow from a store (a time-based event).

Once you have identified these, follow all the data flows and processing that belong to each trigger. All those that can be packaged together compose a function. In a complex system it may not be easy to see the dividing line between one function and another, when an input can trigger a succession of different functions; in SS plc, for example, we see the activities that follow the data flow from a Delegate cancelling a booking, as shown in Fig. 11.2.

Not all of these activities form one function Cancel Delegate Booking. Also included are aspects of Invoicing, maintaining the Waiting List and making Bookings. We cannot say ourselves where one function ends and another begins: it is up to the User to tell us where the line is drawn for any one function. Let us say that in this case the User includes the Processes 7.1, 7.2 in the function Cancel Delegate Booking, and that 7.3, 7.4 and 7.5 belong to separate functions.

It may be that Process 7.2 is performed subsequently in batch mode, in which case we would have to create a temporary data store to provide a trigger for that process. If that were the case, the DFD would look like Fig. 11.3.

Perhaps a function is triggered not by time or by an external flow, but by the combination of such a flow with a particular condition on the database: a prospective delegate on the Waiting List from the same unit, will trigger another Booking. More

144

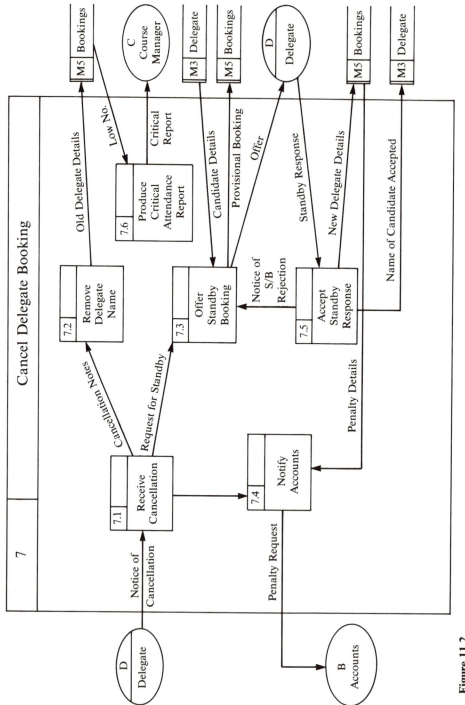

Figure 11.2

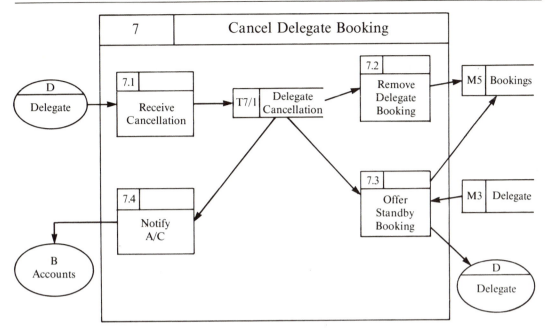

Figure 11.3

drastically, the number of delegates for a scheduled course may fall below an acceptable number and so the Cancel Course process is triggered. In practice, what is more likely to happen in this case is that an exception report will go to the Course Manager who will make that decision.

So we see that in following the life of an input data flow, we must classify each of the subsequent processing activities according to one of our criteria above before deciding if we are dealing with one or more functions. This activity is described more fully below.

At this point we have identified user-initiated functions, and system-initiated functions. Before we leave the DFD and move to our Requirements Catalogue, there is one check to make: that every lowest level process on our Required System DFD has been allocated to at least one function. If not, then something has been missed out; examine the input data flows to that process and identify the function to which it belongs.

The enquiry functions will not be shown on the DFDs. Instead they will be listed in the Requirements Catalogue. Enter these, too, on the Function Catalogue.

Group functions

We have now, with the User's help, identified the basic functions. That is not the end of the task. The User must now go through all the functions with us to identify sequencing of activities, dependencies between activities and the possibility of merging some functions. If one function always includes several subordinate functions, then they can be merged into one large function. Functions 7.3 and 7.5 in Fig. 11.2 could never become just 7.3, as there is a necessary time lag between them. On the other hand, 7.1, 7.2 and 7.4 could be combined into one function, as all are triggered ultimately by the same input flow. If 7.6 is sometimes performed independently of the others, it is kept as a discrete function. As always, it is the User's requirements that dictate how the grouping is achieved.

GROUPING OFFLINE FUNCTIONS

Use the guidelines below to help group together offline functions:

- those events that are triggered by the same external entity, or from related external entities;
- events which are responsible for outputs to a common destination, or related destinations;
- those events which occur simultaneously, or in close succession;
- those events which affect common entities.

The rule to follow here is, if in doubt, leave as separate functions. If they do need to be grouped together after all, it can still be done at Physical Design.

The Users in SS plc have decided that the DFD that shows Cancel Delegate Booking is to be defined as the following functions:

1. Receive Delegate Cancellation, comprising: 7.1, 7.2 and 7.3.
2. Raise Penalty Request, comprising 7.4.
3. Accept Standby Response, comprising 7.5.
4. Produce Critical Attendance Report, comprising 7.6.

Update Function Definitions post-ELH

These functions we have already identified are a part of the input to Entity/Event Analysis in Step 360. For each function that we have identified, an event will be allocated. The normal ratio is one event to one function, but as our grouping may cause a function to be triggered by more than one possible event, that is not a fixed rule.

As Entity/Event Analysis is so much more rigorous than Data Flow Diagramming, (see Chapter 13) it is likely that more events will be discovered than were on the original DFD. Make sure that each of these is allocated to a function. As always, it is the User who will clarify for us how this allocation is to be made, and whether events will be combined for a given function.

Update the Required System DFD and any supporting documentation for any such events discovered here.

Update Function Definitions post-prototyping/Dialogue Design

Possibly the prototyping activities in Step 350 will result in more functions being identified. If this happens these should be given a Function Definition.

Document the functions

In Step 330, as functions are identified, begin a Function Definition form. Figure 11.4 illustrates a form for the SS plc function Cancel Delegate Booking.

Function name
Cancel Delegate Bookings

Function ID
3DG

Function description

This function is occasioned by a Delegate cancelling a Delegate Booking for a given run of a Course Title.
The Delegate details are removed from the Bookings Folder. The Waiting List for that Course Title is examined for other Delegates from the same Branch for ease of billing. Either such a Delegate or else the one waiting longest will be offered a place as Standby.

Error handling

The function is rejected if the system has no record of the Delegate or, if having a record of the Delegate, cannot match it to the Course Run.

DFD processes. 7.1, 7.2, 7.3,

Events: 7.1 Receive Cancellation

Event frequency:
1

I/O Descriptions: D-7.1, 7.1-T7/1, T7/1-m5, T7/1-7.3, 7.3-D, M3-4.3

I/O Structures: 2/1, 2/2

Requirements Catalogue Ref: None

Volumes: Average 15 per week Maximum 30 per week

Related Functions: 4BI, 5ST

Enquiries: None

Common Processing: None

Dialogue Names: Delegate Cancellation (DCA)
Standby Offer (STO)

Service-level requirements:

On-line Response Time	Target:	Range:
	2s	2-7s

Figure 11.4

I/O Description

DFD: Required System

From	To	Data Flow	Content	Comments
D	7.1	Notice of cancellation	Course Code Course Date Delegate No. Branch No. Manager/ID	Input to Process
7.1	T7/1	Cancellation Notice	Course Code Course Date Delegate No. Branch No.	
7.2	M5	Old Delegate Details	Course Code Course Date Delegate Duty	
T7/1	7.3	Request for Standby	Course Code Course Date No. of Places	
7.3	D	Offer	Course Code Course Date Del. Duty Del. Name Del.Address	
M3	7.3	Candidate Details	Branch No. Branch Address Del. Duty Del. Name	

Figure 11.5

I/O Descriptions referenced on this form are identified by the source and recipient of each flow. Events are identified by the process that receives the input data flow that begins the function processing.

Every subsequent function identified during Steps 350 and 360 must be fed back to 330 for creation of this form.

Draw I/O Structures for each function

Every function, whether update or enquiry, has all the data items on inputs and outputs listed on the Function Definition or Requirements Catalogue. If there are enquiries that have not yet had these items specified, they must be identified at this point. Having identified the individual items for the flow, we must identify extra features, such as iterations of groups of data items, optionality, conditionality and so on. Any such features must be noted in the comments column of the I/O Descriptions. Figure 11.5 gives us the description for the I/O to Cancel Delegate Booking.

I/O Structures comprise an I/O Structure Diagram and an I/O Description. The diagram is based on standard structure diagram notation, as in Entity Life Histories and Enquiry Access Paths. The data items are represented as boxes, read in sequence from left to right, as shown in Fig. 11.6.

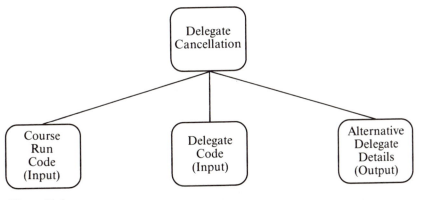

Figure 11.6

Those boxes that act as leaves represent data items or groups of data items that cross the system boundary. Each element should be labelled as input or output. Any groups that repeat are shown as iterations, again with the same notation as ELHs.

If two elements are mutually exclusive, i.e. if one is present the other cannot be, and vice versa, this will be shown as a selection. If an item may or may not be present, that is shown by a selection, one leaf of which will be null. Figure 11.7 illustrates the notation.

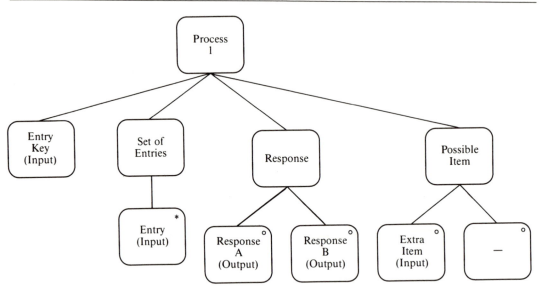

Figure 11.7

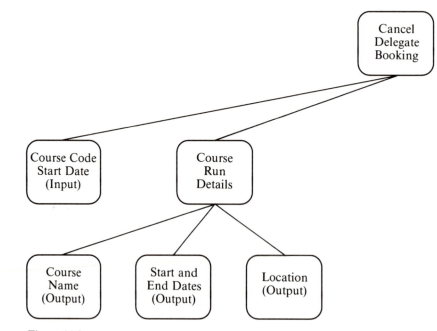

Figure 11.8

OFFLINE I/O STRUCTURES
For each offline function, there will be one input structure and one output structure.

ONLINE I/O STRUCTURES
The online functions are more complex: each structure is likely to contain both input and output elements.

Let us say that an online function Cancel Delegate Booking accepts as input the Course Code and Start Date, i.e. the key for Scheduled Course. The system responds by displaying the full name of the course, start and end dates and the location. This one structure now has input and output elements together, and we have not even executed the function yet! Figure 11.8 shows how this is represented.

So we see that in online I/O Structures, we must interleave those elements that are input with those that are output, in order to represent the full dialogue between User and system. These structures are used as the input for Dialogue Design (see Chapter 14).

The full I/O Structure for this function is shown in Fig. 11.9.

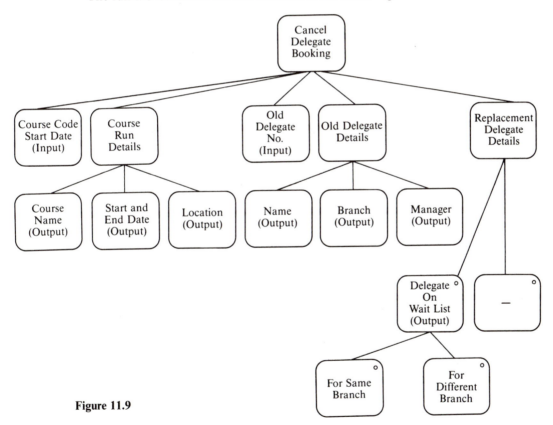

Figure 11.9

11.5   Universal Function Model

Every function in any system can be represented by the model shown in Fig. 11.10. This model is not something created by the practitioner, but is rather a schematic, or checklist, of all that must be specified in a function.

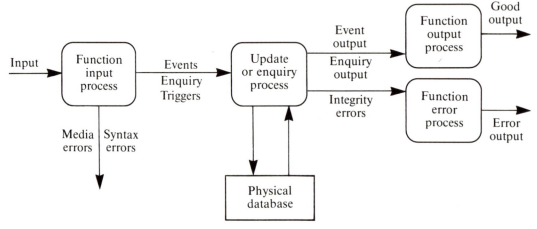

Figure 11.10

The model consists of two basic components: processes and *data streams*. Before the function is completely specified, both components must be defined. The definition is built up in pieces, according to the SSADM stage and techniques being applied. Table 11.1 lists the techniques and the appropriate SSADM products against each component.

The Universal Function Model does not present a model of the Function Definition technique. Rather it provides a conceptual view of a function.

Table 11.1

Component	Technique	SSADM product	Stage
Enquiry trigger	LDM	Enquiry Access Path	3
Event	EEM	Effect Correspondence Diagrams	3
Input & good output	FD	I/O Structures	3
Input & good output	DD	Dialogue Structures	5
Event/enquiry output	LDPD	Process Models	5
Integrity errors	LDPD	Process Models	5
Update processes	LDPD	Process Models	5
Enquiry processes	LDPD	Process Models	5
Syntax & control errors	PPS	Program Specifications	6
Input & good output	PPS	Program Specifications	6
I/O process	PPS	Program Specifications	6

Data streams

The data streams that flow between the processes are defined in terms of data items only, and represent the logical view. Error handling, integrity error messages, control data, including page numbering, current dates, etc. are not represented on the I/O Structures.

The input data is composed of event, enquiry triggers and redundant data, i.e. data that is input with an enquiry, but is not logically necessary for the requirements to be met.

The data items that make up events and enquiry triggers will be defined during Entity/Event Modelling and Logical Data Modelling.

Processes

The processes identified in the model cover the whole cycle of processing from input, through database processing to output processing. In Function Definition we are only concerned with defining the inputs to and outputs from the function. All the stages of process are defined during the specification of the process models in Stage 5, as are the definition of integrity errors.

Syntax errors, control errors and actual processing of inputs and outputs are left until Stage 6, Physical Process Design.

11.6 Supporting documentation

The documentation for the functions is maintained in the Function Definition. There must be one definition for each function as well as at least one I/O Structure. The entries in the definition must record the following information:
- function ID,
- function name,
- User Roles (refer to the external entities on the DFDs),
- Function Description,
- error handling (only in an informal sense—detailed error-handling procedures will be specified elsewhere),
- DFD processes,
- I/O Descriptions (refer to those flows across the system boundary),
- related functions,
- common processing,
- Requirements Catalogue Reference,
- I/O Structures,
- events,
- Enquiry Access Paths (refer to the LDM),
- enquiry elements,
- volumes (numbers of invocations over a period, peak volumes and times),
- dialogue names and
- service-level requirements.

SUMMARY

Function Definition is one of the central features of SSADM. While it is true that SSADM is a data-oriented method, it is always kept in mind that the data is there to support functional requirements, both update and enquiry, online and offline.

We begin identifying functions as soon as we have drawn up the DFD for the Required System, and update the Function Definition with results from Entity/Event Modelling, and from prototyping.

The Function Definitions are carried forward as input to Relational Data Analysis, Entity/Event Modelling, Specification Prototyping, and Logical Database Processing Designs.

12. Relational Data Analysis

12.1 Aims of chapter

By the end of this chapter, you should:
- Understand the concepts and vocabulary associated with Relational Data Analysis (RDA), particularly the nature of the different kinds of identifying key.
- Understand what is meant by normalizing data, and be able to carry out relational analysis to Third Normal Form (TNF); you should also be acquainted with the concepts of Fourth and Fifth Normal Forms.
- Be able to apply the TNF tests to validate the normalized data.
- Know how RDA fits into SSADM, and how it is used with LDM.

12.2 Where RDA is used in SSADM

Relational Data Analysis is used in the following steps.

Step 140   To help with the analysis of the current system, done—if at all here—in conjunction with the LDM, to help identify the data groups and their relationships. If it is used in this context it is usually in an informal manner.

Step 320   To help produce a Required System LDM. At this point we are using the technique to help derive the LDM for the new system. As in Step 140, it is used informally here, if at all.

Step 340   Enhance the Required System LDM. Here the technique is used more rigorously. The RDA procedures are carried out upon the logical data flows to and from functions. An alternative data model to the LDM is drawn up, using the normalized data. This is then used to validate and enhance the Required System LDM.

Use in SSADM

The purpose of RDA is to enhance the Required Data Model. Most of the work in building the model will have been performed before Step 340, but the rigour of RDA, with its bottom-up approach, serves to identify new entities and new relationships.

The source for data to normalize is the set of Input/Output Structures from Function Definition. Only a few selected structures will be subject to normalization.

Inputs to RDA are as follows:
- I/O Structures
- Required System Logical Data Model

Outputs from RDA are as follows:
- Set of normalized relations in at least Third Normal Form
- Enhanced Required System Logical Data Model

Figure 12.1 shows the relationship between RDA and other SSADM techniques.

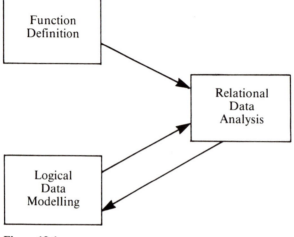

Figure 12.1

12.3   What is RDA?

Like Logical Data Modelling, Relational Data Analysis is concerned with presenting a logical, structured view of data. It is often regarded as being a completely different philosophy from LDM but SSADM uses it rather as a complement. As used in the method it clarifies and enhances the LDM.

RDA derives from mathematical set theory, and comes from the work of Edgar Codd, of IBM (Date, 1986). Codd, among others, recognized that data in computer systems, and indeed in organizations, was gathered and stored in inefficient ways. This was largely because of the ad hoc nature of computerization in the early days of the industry, when computer systems were primarily a way of automating existing manual procedures.

He described a series of steps that formalized and structured data that was previously 'loose', i.e. unstructured, so that data storage was more logical, and independent of specific applications. In this way, update anomalies such as those described in Sec. 12.5 were removed, and data redundancy was reduced.

This chapter does not touch on the theoretical underpinning of this method, but is confined to describing how to apply it to any form of loose data. For those who wish to investigate the theoretical aspects in more depth, the works of Codd (1970, 1974), Date (1986) and Fagin (1983) are recommended.

The purpose of RDA is to reduce, if not altogether eliminate, data redundancy, to resolve update anomalies and ambiguities, to identify all relationships within the data and—in physical terms—to facilitate data maintenance.

The technique involves taking a data source, such as an input form or screen, and 'normalizing' it, (i.e. reducing it successively through a series of steps until it is in the most logical grouping) so that the data on it is in a format known as Third Normal Form (TNF, or 3NF). To reach this stage we have to take that data through First Normal Form (1NF) and Second Normal Form (2NF).

What we obtain from this process is a notation for a logical representation of data, irrespective of physical implementation and storage.

12.4   Concepts and terms

Relation

A relation is a term to describe a table of entries. Figure 12.2 shows us an instance of such a relation, with some entries. The table comprises rows and columns. Each row contains a number of column entries whose sum values make that row unique. Each row is identified by a unique key, or 'true' key. (In Fig. 12.2 the true key is Empno. Emp-name would not be adequate, as names are not unique to an individual.) The remaining columns describe, or qualify, that key. If we compare this to our Logical Data Model, we can call each relation an entity, and each row an occurrence of that entity.

Employee

Empno.	Emp-name	Department	Title	Grade	Mgrno.
A2807	Black	Head-office	Clerk 1	3	C3319
C1166	Jones	Purchasing	Buyer	4	B5834
D4967	Brown	Personnel	Clerk 2	3	A2491

Figure 12.2

The rules of relations, as described by Codd in his paper, include the following:
1. Each row must be unique, i.e. duplication of keys is not permitted.
2. There must be no significance in the order of rows.
3. There must be no significance in the order of columns.
4. Each column (attribute) may only have one value per row.

Codd has prescribed many more rules to define and limit relations since the earlier papers, but these will serve us for the purposes of this chapter.

Key

A key is a column or group of columns that uniquely identifies a given row in a relation. The terms *true key* or *determinant* are sometimes used: the terms key, true key and determinant are synonymous.

Cust. No:			Order No.	

Cust. Name:

Cust. Address:

Cat. No.	Description	Price	Qty

Figure 12.3

SIMPLE KEY

This is a single column that identifies a row. Part Number, Employee Number, NI Number are typical examples of simple keys to retrieve a given part, employee or claimant.

COMPOUND KEY

Sometimes a single column is not sufficient to identify a row, and a combination (or *concatenation*) of columns is required. For example, Fig. 12.3 shows a form for a customer order. We see that it is divided into a header, with an order number and customer details, and entries, which give the details of each stock item ordered.

If we need to access this customer order at any time we need a key. This is easy: every new order that comes into our system is automatically given a unique Order No. To retrieve details of the whole order, the Order No. is a sufficient key.

The individual entries, however, need more information than that to identify them. It may be that one order is for 50 separate stock items. As we are only allowed one value per column entry on a row in a relation, we must treat these order entries differently from the whole order. These entries will form a separate relation (Fig. 12.4).

Cat. No.	Description	Price	Qty

Figure 12.4

To access a given entry on a customer order we must identify a unique key for that entry. We want to identify an entry on a given order, so Order No. would be part of the key; we also want a specific item on that order, so we add the Cat. No., giving us a concatenated, or *compound*, key.

Cat. No. is chosen as the other part of that key, because in this case it is unique to each entry on the order. Other columns such as Description or Quantity may not be enough to identify one entry only, without ambiguity or mistake; Cat. No. is sufficient.

COMPOSITE KEY

A composite key is, like the compound key, a concatenated key, but there is a crucial difference. Each element of a compound key is unique: there is only one of each Order No. in our system, and there is only one of each Cat. No. in the system.

One element of a composite key, though, is not unique in the system, and has to be qualified by another element.

In the above example of the Customer Order, the key element inside order is Cat. No. However, each entry on the Customer Order takes up one line on the order form. This gives us another unique identifier for each entry: the Line No.

Thus, if we were to take the key element as Line No., as in Order 5638/90 Line 1, Line 2, Line 3, etc., we see that the system will have many instances of Line 1, Line 2, and so on. It is not helpful to us then to ask for details of Line No. 10 in our order system. We will be offered possibly 5000 entries called Line No. 10 to sift through. To identify the one we want uniquely we must marry it with the order, as in Order No. 5638/90 Line No. 10.

The difference between composite and compound keys is that all elements in compound keys are unique, but in composite keys, one is not unique. In the above example, the compound key would be Order No./Cat. No.: 5638/90–015699/A. The composite key is Order No./Line No.: 5638/90–10. Either would be sufficient, though, to identify that particular entry on that particular order.

Another example of a composite key is the identifier for a machine on a factory floor. A machine may be a cutter, a drill, a lathe or whatever. Each *type* of machine will have its own code. For example, a lathe for turning steel may have the code Ltst.

In our factory, there may be six such lathes, each one needing to be uniquely identified. As each machine comes into service, it is given a serial number: 1, 2, 3, 4, etc.

No. 1, No. 2 and No. 3 by themselves have no meaning in our system: we have a dozen machines with the serial no. 3. What we need is information about the machine that has serial no. 3 *and* turns steel. In this case, the key would be Ltst/3.

CANDIDATE KEY

A candidate key is any possible unique identifier for a relation. Our key will be chosen, be it simple, compound or composite. Others which would serve equally well are candidate keys. They are not in fact keys unless they have been chosen as such. Where a relation has two or more candidate keys, the one selected may be an arbitrary choice.

In SSADM, every determinant should, at the end of the analysis, be examined to make sure that it is in fact a candidate key. If we take the example of a seating plan for a course, we could find the relation below.

Course code
Course date
Room-no.
Seat-no.
Delegate name

This is an acceptable relation, with a valid key, but it is not the only possible key. An equally valid key would be:

Course code
Course date
Delegate Name

The attributes for this key would then be Room-no. and Seat-no., giving us an equally valid, and different, relation. Accordingly, we must include this in our set of relations.

Take care that you only do this with candidate keys that give a new relation, rather than with alternative keys that give the same relation, e.g.

Pay No.
NI No.

could equally well have been expressed as:

NI No.
Pay No.

There is no new information here, and so no new relation.

FOREIGN KEY

A foreign key is a data item in one relation, and a primary, or true key in another. The item is referred to as a foreign key in that relation in which it occurs as an ordinary data item. This is a mechanism whereby relationships between data groups can be implemented.

Thus, in Figure 12.5, an Employee relation, we have the attribute Projno. to show on which project the employee is working. In the same system, we will have a relation Project, whose key is Projno., the same attribute. In the Employee relation, the attribute Projno. is the foreign key, linking a specific employee with a specific project.

Conversely, an attribute of Project is Mgrno., to identify the project manager. Further investigation tells us that the value for Mgrno. is taken from the same domain as Empno. Therefore, Mgrno. in Project is also a foreign key, linking the relation to Employee, but in a different relationship.

Employee

Empno.	Surname	Forename	Grade	Projno.*	Start-date
Z0/251	Green	Marion	4	Books90	12/10/90
J0/668	Ashe	Geoffrey	3	Timet90	10/12/90

Project

Projno.	Name	Start-date	End-date	Mgrno.*
Books90	Bookings	12/10/90	13/06/91	H3/887
Timet90	Timetable	10/12/90	01/04/91	T6/115

Figure 12.5

Dependency

When we say that data item A is *dependent* on data item B, we mean that we need to know the value for B to know the value for A. In the example in Fig. 12.4, Start-date and End-dates are dependent on Projno. That is, given a value for Projno., we can find the values for Start-date and End-date.

When a key is identified, the dependency is self-evident; the use of the idea is in identifying the key to begin with, especially when moving to Second and Third Normal Forms.

12.5   How to carry out RDA

Unnormalized form

The first task is to express all data from the source in an *unnormalized* format, i.e. to list all the data items and place them under a suitable (unique) key. This format is termed a *raw relation*.

Let us take a form from an employment agency called Jobs-For-U (JFU). This particular form is a registration form for new clients, and is filled in by job-seekers when they sign on to JFU's books. It is called a job-seeker's registration form, and is illustrated in Fig. 12.6.

This form presents us with a number of data items about our job-seeker, but these are not in a logical grouping. In fact, the data given to us tells us about rather more than the job-seeker, and so we apply the rules of normalization to give us that more logical grouping.

The first task is to lay out this data as a list of data items, in what is called Unnormalized Form (UNF). This comprises the list of data items (attributes, or fields) under a unique, identifying key. Selecting the key in this instance is a simple task, because the form itself provides us with a unique key: Job-Seeker Registration Number, or Jsregno. We identify the key field by underlining it. So our first pass at UNF will be as shown in Fig. 12.7. Note:

1.   Two sets of data items can have multiple values on this form: those to do with qualifications, and those to do with previous jobs. Those groups of data items are indented under an umbrella name for each. The first umbrella name is Js-qualifi-cations. The second is Previous Employment.
2.   Each job-seeker is allocated a Job-code to describe the nature of work most suitable, e.g. Machine Fitter, Invoicing Clerk, Van Driver, Clothes Shop Assist-ant, etc. They are also allocated a second such code, called Alternative-code.

Our second pass at UNF involves listing separately those repeating groups and choosing a key for each group. This gives us Fig. 12.8.

Job-Seeker Registration	No.			
Name:	Age:			
Address:	Date of Birth:			
	Job Code:			
	Alternative Code:			
Tel:				

Qualifications:

Subject	Level	Year	Result	Awarding Body

Previous Employment:

Employer Name:	Job Title	Date started	Date left
Employer Address:			
Telephone:			
Reason left:		Pay:	
Employer Name:	Job Title	Date started	Date left
Employer Address:			
Telephone:			
Reason left:		Pay:	
Employer Name:	Job Title	Date started	Date left
Employer Address:			
Telephone:			
Reason left:		Pay:	

Registration Officer:

Duty Code: _____

Name: _____   Telephone no. _____

Figure 12.6

UNF

<u>Jsregno.</u>
Js-name
Js-address
Js-telno.
Js-age
Js-DoB
Js-qualifications:
   Subject
   Level
   Awarding Body
   Year Awarded
   Grade/Class
Previous Employment:
   Employer
   Employer-Address
   Employer-Telno.
   Job-title
   Date-started
   Date-left
   Reason-left
   Pay
Job-code
Alternative-code
Registering-officer-Duty
Registering-officer Name
Registering-officer Telno.

Figure 12.7

Job-seeker
   <u>Jsregno.</u>
   Js-name
   Js-address
   Js-telno.
   Js-DoB
   Job-code1
   Alternative-code
   Registering-officer-duty (R/O)
   R/O Name
   R/O Telno.
Js-qualifications
   <u>Subject</u>
   <u>Level</u>
   Year-awarded
   Awarding Body
   Grade/Class

Employment History
   <u>Employer</u>
   <u>Job-title</u>
   <u>Date-started</u>
   Employer-address
   Employer-telno.
   Date-left
   Reason-left
   Pay

Figure 12.8

Each repeating group in this example has a compound key of several elements.

Js-qualifications has a key of Subject (English, Maths, Statistics, Business Studies) and Level (CSE, O-level, A-level, HND, M.Sc., etc). This is because Subject by itself is neither unique nor sufficiently informative: a Job-seeker may have O-level Maths, and A-level Maths. The value 'Maths' would not extract all the qualifications. 'Maths' and 'A-level' would.

If an employer were looking for someone with a business studies degree, extracting the Job-seeker and Subject alone would not be sufficient. Excluding Level from the key would give him all of the CSE and O-levels as well.

Employment History has a key of Employer, Job-title and Date-started. For the following reasons, the key needs to contain all of these elements to give each entry a unique value. Perhaps a Job-seeker has worked for the one employer on different occasions in different jobs. Employer by itself would not be enough, then, to identify each entry.

Even Employer, Job-title, is not enough: it is possible, after all, that one person has worked as a machine fitter at a plant, left or been laid off, and then returned to the same plant as a machine fitter again, some months later. Therefore that person could have left a job as machine fitter for the one plant twice, giving us two possible values for that data item. Date-started as part of the key will tell us which of those two values is the appropriate one. This gives us Unnormalized Form. Figure 12.13, later in the chapter shows how this is entered on SSADM forms.

First Normal Form

First Normal Form identifies a relationship between groups of data items that corresponds to one-to-many relationships on a Logical Data Structure, as in Fig. 12.9.

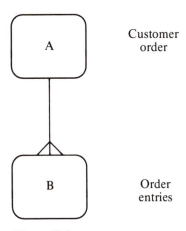

Customer order

Order entries

Figure 12.9

The procedure for taking the data from Unnormalized Form to First Normal Form is to examine all repeating groups, find their own unique key and make them separate relations.

The key for repeating groups will be a compound key, made up of the key for the original UNF (to preserve the one-to-many relationship) and one or more other elements that will uniquely identify each occurrence of the repeating group.

Applying this 1NF rule, then, gives us Fig. 12.10.

First Normal Form (1NF)

Job-seeker
    <u>Jsregno.</u>
    Js-name
    Js-address
    Js-telno.
    Js-DoB
    Job-code1
    Alternative-code
    Registering-officer-duty (R/O)
    R/O Name
    R/O Telno.

Js-qualifications
    <u>Jsregno.</u>
    <u>Subject</u>
    <u>Level</u>
    Year-awarded
    Awarding body
    Grade/Class

Employment History
    <u>Jsregno.</u>
    <u>Employer</u>
    <u>Job-title</u>
    <u>Date-started</u>
    Employer-Address
    Employer-Telno.
    Date-Left
    Reason-Left
    Pay

Figure 12.10

Second Normal Form

WHY

The result of putting the data into 1NF means that some data items are identified by a key, without necessarily being dependent on the whole key. Thus, in the new relation, Employment History, we have the attribute Employer-address dependent on four key items, including Job-title and Jsregno.

This is unacceptable: if we want to remove references to a given Job-seeker, we risk losing information about an employer on our lists. Conversely, if over 100 of our clients have worked for the one employer during their history, following mass redundancies from one plant, for example, then for each one of those Job-seekers, we would store the address of the same employer. To avoid such anomalies, we decompose our relation further to Second Normal Form (2NF).

To achieve 2NF we examine all relations with compound keys for 'part key dependencies', i.e. does each attribute need the *whole* key to identify a value, or only part of the key. In the example just quoted, Employer-address depends only on Employer to identify its value. Date-left, on the other hand, needs the entire key for us to identify a value.

HOW

When going from a 1NF relation to a 2NF relation, copy the entire compound key across first. It may be that one or more data items will depend on the whole key, or it may be that all depend only on part. In either event, the whole compound key must be in a 2NF relation. The key itself has important information connecting, in this case, Job-seeker with Employer with Job-title that must not be lost.

For every other data item in the relation, ask the following question: to obtain one and only one value for this data item, do I need a value for the whole key, or just a part of it?

Second Normal Form (2NF)

Job-seeker
  <u>Jsregno.</u>
  Js-name
  Js-address
  Js-telno.
  Js-DoB
  Job-code1
  Alternative-code
  Registering-officer-duty (R/O)
  R/O Name
  R/O Telno.

Js-qualifications
  <u>Jsregno.</u>
  <u>Subject</u>
  <u>Level</u>
  Year awarded
  Awarding Body
  Grade/Class

Employment History
  <u>Jsregno.</u>
  <u>Employer</u>
  <u>Job-title</u>
  <u>Date-Started</u>
  Date-Left
  Reason-Left
  Pay

Employer Details
  <u>Employer</u>
  Employer-address
  Employer-telno.

Figure 12.11

In the relation Js-qualifications, the answer for each of our data items is 'We need a value for the *whole* key'.

In the relation Employment History the answer for Employer-address and Employer-telno. depends only on a value for Employer No., not on Date-started or Jsregno. Therefore it is not yet in Second Normal Form.

All other data items in the relation depend on the whole key for a value, so we leave them as they are. We can decompose our 1NF relations, then, as shown in Fig. 12.11.

If we now remove reference to one of our job-seekers, we do not lose any information about an Employer's address or telephone number. Conversely, if an Employer moves premises or changes telephone number, we only need to record that change once, rather than on every related Job-seeker's record.

Third Normal Form (TNF)

WHY

So far, we have eliminated several storage and update anomalies in our passage to 2NF, but there are still more.

If we return to our relation Job-seeker, we find that on every registration form we hold details of the Registering Officer. It is right that we identify the officer responsible for registering and classifying each Job-seeker, but details such as name and telephone number do not need to be stored each time, and deleted each time we remove reference to a Job-seeker.

Job-seeker
   <u>Jsregno.</u>
   Js-name
   Js-address
   Js-telno.
   Js-DoB
   Job-code1
   Alternative-code
   Registering-officer-duty (R/O)
   R/O Name
   R/O Telno.

Duty-no.	Surname	Forename	Section	Date Started
IN77	Black	Robert	Interview	20.8.87
RE34	Patel	Kaushik	Registry	15.10.86
VA12	Green	Marion	Vacancy	30.5.89
IN50	Patel	Amresh	Interview	12.3.90

Figure 12.12

HOW

The move to Third Normal Form (TNF) removes this anomaly by examining dependencies between data items. Thus in our example, we examine the data in the relation Job-seeker, asking the following for each pair of items: given a value for data item (a), is there one and only one possible value for (b)? Let us look at the Job-seeker relation again to illustrate this. To help us recognize the dependencies, Fig. 12.12 includes a table for JFU employees.

Given a value RE34 for Duty-no., can we provide one name and one name only? Assuming Duty-no. to be a unique key identifying a person, the answer clearly is 'yes'. If we ask the question 'Given a name Patel, can we provide one Duty-no. and one only?' the answer is 'no', there are two Patels working there, each with a different Duty-no.

To confirm this inter-data dependency, we ask the question, 'Does R/O-Name depend for a value directly on Js-Regno, or directly on Duty-no.?' The answer is clearly, 'Duty-no', so we have truly found an inter-data dependency.

Now that we have identified that dependency, we can extract the data items that describe the R/O and assign them to their own key. The actual key of the new relation (in this case R/O Duty-no.) must stay with the original relation, and becomes a foreign key. Thus we can still identify which R/O is responsible for the Job-seeker.

In 3NF, then, we can decompose our relation Job-seeker to the relations in Fig. 12.13.

Job-seeker
   **Jsregno.
    Js-name
    Js-address
    Js-telno.
    Js-DoB
    Job-code1
    Alternative-code
   *R/O Duty
Job-seeker/RO
   **R/O-duty
    R/O-name
    R/O-telno.

Figure 12.13

On examining all of our relations at 2NF, we see that each data item depends upon the whole key, and only the key, so they are already in Third Normal Form.

Figure 12.14 shows us the complete process from UNF to 3NF. All of this information came only from the single document Job-seeker Registration Form.

12.6 Optimization

The example worked in Fig. 12.14 was taken from just one source document. However, when carrying out the analysis and design of a system, we encounter many sources; generally they are the I/O Structures/Descriptions. Data items will occur several times through the system and this process will identify a number of relations that share the same key.

One of the objectives of RDA is to rationalize data storage and maintenance. The

UNF	1NF	2NF	3NF
<u>Jsregno.</u>	<u>Jsregno.</u>		<u>Jsregno.</u>
Js-name	Js-name		Js-name
Js-address	Js-address		Js-address
Js-telno.	Js-telno.		Js-telno.
Js-age	Js-age		Js-age
Js-DoB	Js-DoB		Js-DoB
Js-qualifs	Job-code		Job-code
Subject	Alt. Code		Alt. Code
Level	Reg. Officer		*Reg. Duty Code
Award. Body	Duty-code		
Year Award.	Name		<u>Reg. Duty Code</u>
Grade/Class	Telno.		Name
Prev. Empt			Telno.
Employer	Js-qualifs		
Emp. Add.	<u>Jsregno.</u>		
Emptelno.	<u>Subject</u>		
Job-title	<u>Level</u>		
Date-start	Award Body		
Date-left	Year Award.		
Reason	Grade/Class		
Pay			
Job-code	Prev. Empt	Prev. Empt	
Alt.Code	<u>Jsregno.</u>	<u>Jsregno.</u>	
Reg. Officer;	<u>Employer</u>	<u>Employer</u>	
Duty-code	<u>Job-title</u>	<u>Job-title</u>	
Name	<u>Date-start</u>	<u>Date-start</u>	
Telno.	Date-left	Date-Left	
	Reason	Reason	
	Pay	Pay	
	Emp. Add.		
	Emp. Telno.	Employer	
		<u>Employer</u>	
		Emp. Add.	
		Emp. Tel.	

Fig. 12.14

next step is to examine all relations, and merge those that share a key, so that a true key is the key of one relation only. The following steps will achieve this.

1. Examine all the data items for synonyms and homonyms, i.e. in one relation we may have a key Cust-no., and in another, Account-no. Further investigation reveals that both fields describe the customer identifier, but different business functions use different terms. Decide which term you will use and be sure that you always use it in your data model. That is an example of a synonym, two terms describing the one item.

    You may then discover that you have two items called Account-no.; one is the customer identifier, as we have seen, and another refers to a financial account held by the company. Here we have a homonym, one term for two discrete items.

2. Having satisfied yourself that all synonyms and homonyms are identified and dealt with, identify all relations that share a key. Merge them so that all of the attributes form one relation.

3. Examine the merged relations to ensure that they are all in Third Normal Form. It is possible that in the merging of relations you have created new inter-data dependencies (or transitive dependencies).

Now you have identified all of the data items in the system, and upon which key each one depends. Having carried out this very detailed analysis of the data, you will convert it into a data model, so that with the LDS you can create the logical model that will provide the basis for the physical database design. This will be the subject of Sec. 12.7.

12.7 Enhance Required System LDM

LDM gives us a view of the system data that begins with entities and relationships, and works down to data items via the supporting documentation. With RDA, we begin at the lowest level with the data items and their determinants, and work up to the broader view of entities and relationships. There is a simple five-step algorithm to achieve this.

After we have optimized the relations, and confirmed that they are still in Third Normal Form, we perform the five steps (Note: the terms 'entity' and 'relation', below, are interchangeable.):

1. Represent each relation as an entity, i.e. draw a box and name it after the relation.

2. If the primary key is a composite key, mark the part that is unique, i.e. the 'qualifying' element, as a foreign key.

3. Make sure that all masters of compound key relations are present. This means that every element in a compound key must be the primary key of another entity. If such an element is not the key of another entity, then create one, with that element as the key.

   Any other entity that contains that data element as part of a compound key will also be a detail of this new entity. This element will be marked as a foreign key in all relations where it appears as a non-key data item.

4. Make compound key relations into details. Where an entity has a compound key, it will be a detail of all entities that has one of the key elements as its entire key. This is a simple extension of rule 3 above.

5. Make relations with foreign keys into details. Every data item that is marked as a foreign key is, by definition, a primary key of another relation. That other relation will now be shown as a master of the relation with the item marked as a foreign key.

   Figures 12.20 to 12.24 illustrate these steps.

The application of the rules above create another LDM which is similar to the original LDM. It is not intended to supersede the original, but to enhance it. If the functional requirements demand features on the relational model to be included on the data model, then include them. If the new entities forced out by RDA do not serve any functional requirements, do not include them.

12.8   Super Systems plc

This section demonstrates the whole process of normalizing to Third Normal Form, and building a relational model using the five rules. The source for this is the SS plc I/O Structure for the function: Make Booking. The structure is shown in Fig. 12.15. Boxes marked 'Details' contain extra data items: the supporting I/O Structure Description is the source for all data items, which will find their way onto UNF.

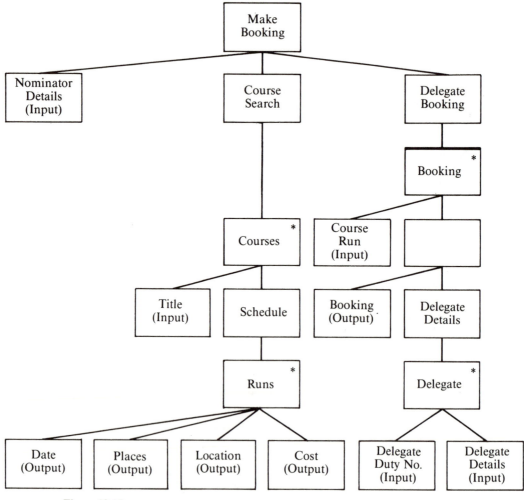

Figure 12.15

Unnormalized Form appears in Fig. 12.16.

Make Booking

<u>Nominator Duty-code</u>
Nominator Address
Nominator Phone

<u>Course Title</u>
<u>Course Date</u>
Max. Places
No. Places Spare
Location
Cost

<u>Booking No.</u>
<u>Delegate Duty Code</u>
Delegate Name
Delegate Phone
Course Title
Course Date
Special Factor

Figure 12.16

To reach 1NF we expand the key of the repeating groups, in this case Course Title and Delegate. First Normal Form appears as in Fig. 12.17.

<u>Nominator Duty-code</u>
Nominator Address
Nominator Phone
Booking No.

<u>Nominator Duty-code</u>
<u>Course Title</u>
<u>Course Date</u>
Max. Places
No. Places Spare
Location
Cost

<u>Nominator Duty-code</u>
<u>Course Title</u>
<u>Course Date</u>
<u>Booking No.</u>
<u>Delegate Duty Code</u>
Delegate Name
Delegate Phone
Special Factor

Figure 12.17

We carry out the part-key interrogation for each relation with a compound key to give us Second Normal Form, as in Fig. 12.18.

Nominator Duty-code
Nominator Name
Nominator Address
Nominator Phone

Nominator Duty-code
Course Title
Course Date
Booking No.

Course Title
Course Date
Max. Places
No. Places Spare
*Location
Cost

Figure 12.18

Nominating Officer

Nominator Duty-code
Nominator Name
Nominator Address
Nominator Phone

Booking

Booking No.
*(Course Title
Course Date)
*Nominator Duty-code

Course Run

Course Title
Course Date
Max. Places
No. Places Spare
*Location

Course Title

Course Title
Cost

Course Location

Course Title
Location
Max. Places

Delegate

Delegate Duty Code
Delegate Name
Delegate Phone Number
Special Factors

Delegate Booking

Booking No.
Delegate Duty Code
*Nominator Duty-code

Figure 12.19

Moving to Third Normal Form we examine the 2NF relations for inter-data dependencies. Remember that we are examining not only the non-key data items for dependencies, but also the components of compound keys as well. The aim is to identify *all* possible compound keys, to make them determinants in their own right. This can be the most complex part of the RDA procedures, as every set of data items must be examined.

In the example shown in Fig. 12.19, we find that after first examining compound key items, and non-key items, there is still a determinant hidden in one relation: Course Title may have a different set of Max. Places, depending on which location is being used, so another key emerges: Course Title/Location. This is not immediately obvious on the first pass through 2NF relations, but when such a case is identified, a new relation must be created in addition to the existing relation. This example is significant: it will be used in scheduling courses. (This process is known as *Boyce–Codd Normal Form*. It is a refinement of Third Normal Form, rather than an addition to it, and is intended to fill any holes that straightforward application of TNF procedures might leave.)

We now apply the five conversion rules to the TNF relations to create the RDA model.

1. Make each TNF relation into an entity (see Fig. 12.20).

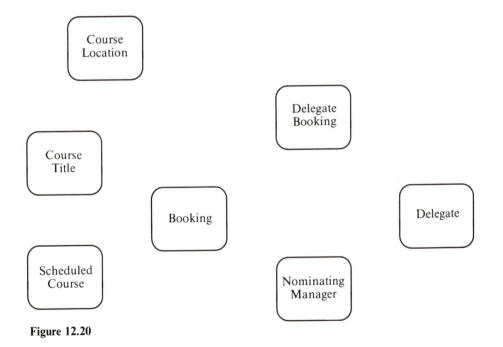

Figure 12.20

2.   Mark qualifying elements of hierarchic keys as foreign keys (see Fig. 12.21).

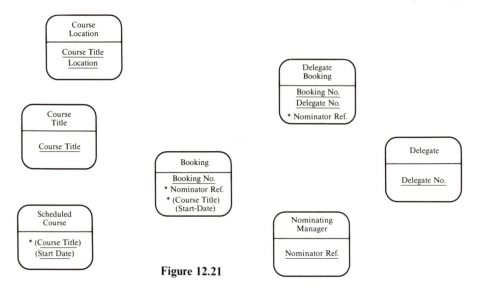

Figure 12.21

3.   Make sure that all masters of compound keys are present (see Fig. 12.22).

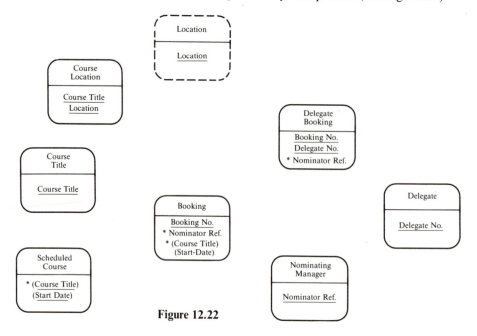

Figure 12.22

4. Make compound key relations into details (see Fig. 12.23).

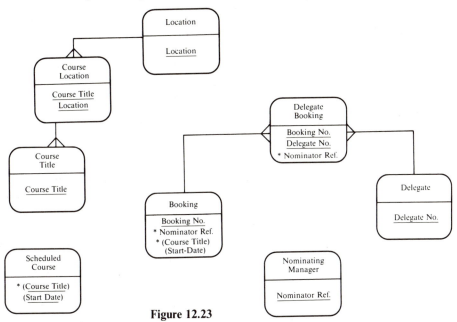

Figure 12.23

5. Make relations with foreign keys into details (see Fig. 12.24).

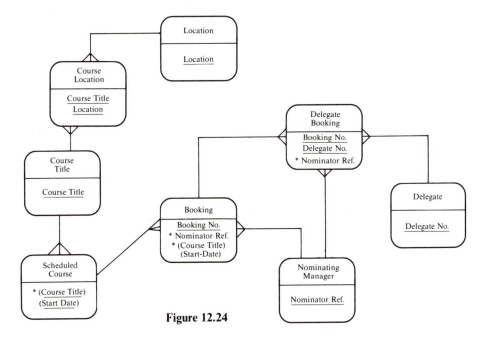

Figure 12.24

12.9   Beyond Third Normal Form

Sometimes 3NF is not rigorous enough to extract all of the dependencies between data. There are two further levels of normalization available, Fourth and Fifth Normal Forms. As these are rare occurrences—especially Fifth Normal Form—I shall not go into too great detail here. Readers who want to explore these further are referred to Date (1986).

Fourth Normal Form

This case is usually found if an unsuitable primary key is chosen at UNF or 1NF. We leave the notion of functional dependency here, and instead look at what is called *multivalue dependency* or MVD.

MVD is another form of the one-to-many relationship; a single value of a given data item may have multiple values of another data item associated with it. This is likely to happen in compound-key only relations; any other sort of relation would have removed such dependencies earlier in normalization.

MVD means that there is a $1 : m$ relationship between data items that was not resolved at 1NF. To resolve such a situation, another relation must be created, with the 1 element as the determinant.

In SS plc, for example, we may have a situation where certain specialisms are needed to teach on certain courses, and lecturers possess particular specialisms. 3NF analysis may have led us to the key-only relation of Fig. 12.25. There are MVDs between Lecturers and Specialisms, and Course Types and Specialisms. To resolve these, we create two new relations, as in Fig. 12.26. These are now in Fourth Normal Form (4NF).

Lecturer Code
Course Code
Specialism

Figure 12.25

Lecturer Code   Specialism
Specialism   Course Code

Figure 12.26

Fifth Normal Form

Fifth Normal Form has been described by Date as a pathological case because it is so rare that decomposition at this level is needed. As with 4NF, we are dealing with key-only relations, and with keys that contain three or more elements. It may be that they cannot be broken down into two further relations, but can—and therefore must—be broken into three.

There are no examples from SS plc to demonstrate this. If we return to JFU, from Sec. 12.5, we can take an example of a Job-title such as Driver, Fitter, Counter Assistant, etc., which is common to several employers. Several job-seekers may have several different job-titles at different employers.

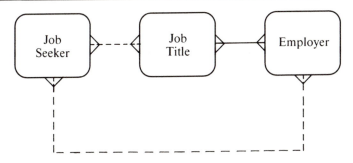

Figure 12.27

The LDS for such a situation is as in Fig. 12.27. The key-only relation resolving this would be:

Jsregno.
Employer
Job-title

To split this into two relations:

Jsregno	Employer
Job-title	Job-title

may give us spurious relationship, suggesting that a specific Job-seeker and Employer are joined by a job-title, when in fact they are not related at all.

To resolve this, without losing information, we have to create three new relations, rather than two. The relations would be:

Jsregno.	Job-title	Employer
Job-title	Employer	Jsregno.

Figure 12.28 shows the corresponding LDS with these relations.

These relations are now in 5NF. It is very unlikely that you will ever meet such a case in a real-world system, and need to model it. If you should, it may be a sign of muddled analysis earlier on in the Requirements Specification procedures, and you should turn your attention to that area first.

I have avoided giving any formal mathematical definitions or proofs of the different Normal Forms in this chapter; such proofs are outside the scope and intentions of this book. If you wish to study the formal mathematics that underlie the normalization process, Date (1985) describes it in considerable detail.

SUMMARY
Relational Data Analysis is intended as a method of identifying the true key of every data item in the system. Using this approach, we expect to reduce data redundancy to a minimum, and produce a model that is flexible and easily maintained.

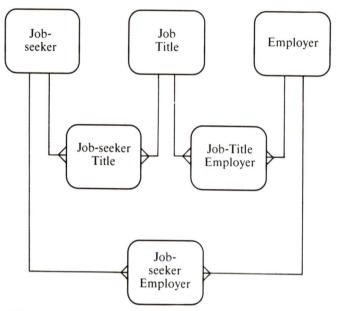

Figure 12.28

Because it is based on a mathematical principle, it is rule-based, and so relatively simple to apply, both for the novice and the experienced analyst.

Another feature of RDA that makes it fit the philosophy of SSADM is its independence from any physical implementation chosen. While it obviously suits a relational database approach, it can equally be implemented on a Codasyl database, or hierarchical, or even conventional flat-file structures. Implementation considerations, therefore, neither dictate nor are constrained by the RDA.

1. *UNF→1NF*   We identify the key upon which each data item/group depends: by separating out repeating groups we identify for each data item *the key*.
2. *1NF→2NF*   We identify which part of compound keys govern each data item: by identifying part-key dependencies we find for each data item *the whole key*.
3. *2NF→3NF*   We look for dependencies between data items rather than between data item and key: by examining such transitive dependencies we ensure that each item depends on *nothing but the key*.

EXERCISE

Below is a set of data items relating to a hospital ward. Carry out Relational Data Analysis to 3NF or 4NF, and produce a relational model.

Ward Rota (by day)

Ward Name
Date
Ward Type
No. of Beds
Nurse Personnel No.
Nurse Name
Nurse Experience
Patient No.
Patient Name
Bed No.
Admission Date
Expected Discharge Date
Ward Sister

Consultant's Diary

Consultant
Date
Specialism
Patient No.
Patient Name
Ward Name
Bed No.
Sex

Assumptions: A patient may be moved to a different ward and/or bed during his or her stay. Sister and Nurse No. are the same code.

13. Entity/Event Modelling

13.1   Aims of chapter

In this chapter, you will learn:
- The use of Entity Life Histories in SSADM.
- The differences between entities, events and effects.
- How to construct the Entity/Event Matrix.
- How to draw and review the ELH chart.
- How to add the operations to the chart.
- How to draw a supporting Effect Correspondence Diagram.
- How to add the state indicators to the chart.

13.2   Where Entity/Event Modelling is used in SSADM

SSADM creates and amends ELHs in the following steps of analysis and design:
- *Step 360*   Develop Process Specification
- *Step 520*   Define Update Processing Model

Use in SSADM

Entity/Event Analysis presents us with our third view of the data. DFDs and LDSs give us a snapshot of their views of the data; ELHs provide us with the dynamic element of time's action on the system, and so enable us to show (and define) how each entity is updated within the system, whether that update be creation, modification or deletion.

Inputs to Entity/Event Modelling are as follows:
- Data Catalogue
- Function Definition
- I/O Structure
- Logical Data Store/Entity Cross Reference
- Required System LDM
- Requirements Catalogue

Outputs from Entity/Event Modelling are as follows:
- Entity Life Histories
- Effect Correspondence Diagrams

Figure 13.1 illustrates the relationship of Entity/Event Modelling with other SSADM techniques.

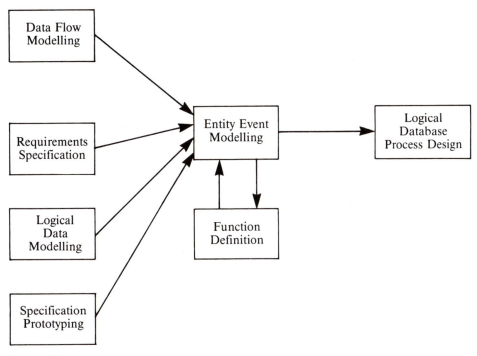

Figure 13.1

ELHs in analysis

In Stage 3 ELHs provide a validation for DFD and LDS, enabling us to verify their completeness and correctness. On the one hand we can see how each entity on the LDS is created, modified and deleted, and on the other we can confirm that the updating processes are connected to the relevant data stores on our DFD.

As described in Chapter 8, ELH is one of the techniques to use the structure diagram notation of sequence, selection and iteration.

In Stage 3, the events that affect each identity are recorded, as well as the business rules which govern the permitted sequence of activity. Operations are added according to the roles of the entity as defined by the processing of the event.

After the ELHs are drawn, an Effect Correspondence Diagram is created for each event. This will be carried forward into Logical Database Process Design.

ELH in design

In Stage 5, state indicators are added to the ELHs, to allow semantic validation to take place (i.e. is what is represented on the model a true reflection of what happens in the world?). The Effect Correspondence Diagrams are transformed into Update Processing structures, one per event.

13.3   Definition of terms

Entities

An entity, as defined in the LDS and described in ELH is a generic view of 'a thing about which we hold information'. Delegate is the generic entity in SS plc; Arnold Laine of Solihull is a specific *occurrence* of that entity. (It can be seen that as we talk of occurrences, we think in more physical terms than when we speak of entities, and it will help us therefore to think of each *occurrence* as a record.)

The ELH is a comprehensive picture of all that may happen to any one of our occurrences from the time that the record is created to the time of its deletion. Some occurrences will have straightforward, uncomplicated 'life histories', while others will experience anomalies, e.g. a customer may be declared bankrupt before an order of his is processed, and therefore all details of his order come to an unexpected halt. The ELH for this entity type must cater for such anomalies, as well as describing the standard 'quiet life'.

Events

An event is something that happens in the real world to cause a change to be made to the database. In our model, it causes updates to one or more entities. We usually, though not always, show the trigger as a data flow across the system boundary. 'Event' does not refer to the processing itself, but to that in the real world which causes the processing.

Events are reckoned to be of three kinds:
1.   Externally generated, such as a Nominating manager placing bookings for staff to attend a course. This will be shown as a flow across the boundary.
2.   Internally recognized, such as Course Numbers falling below a critical level, so cancelling a run of a course. This will have an initial trigger of a flow across a boundary, but that itself is not sufficient to be called an internally recognized event; for that, we need the database to be in a particular condition, and so on the DFD we also see a flow coming from a data store into the process.
3.   Time-based, such as the automatic raising of invoices at a fixed time each month. This will be shown as a flow going from a data store to a process as the sole trigger to that process.

When naming events, do not copy the name of the DFD process that carries out the update; it may be that more than one trigger can activate that process, each trigger being a different event. By naming an event after a process, a cause of that process may be missed.

Effects

The changes in an entity are the effect of an event. The leaves of the chart reflect the events, and must be named after the event.

Sometimes, an event may have different effects on an entity, the effects being mutually exclusive. This can happen, for instance, in SS plc where a delegate is booked on a Course Run. If that delegate is already on the Waiting List for the

Course Type, one effect will be to disconnect the Delegate Booking from that relationship. If, on the other hand, the booking is straight from an application on to a Course Run, that will not apply.

Where this may happen, show the possible effects as a selection on the diagram, and in parentheses name the possible roles. Figure 13.2 illustrates this.

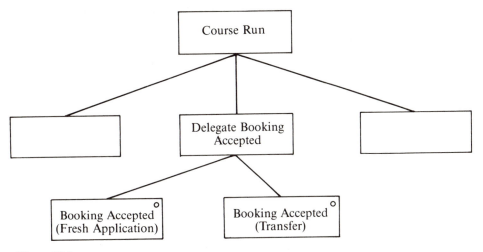

Figure 13.2

The effects noted at each leaf are modifications to the database, be the modification the creation of a record, a change to one or more of its fields, or its deletion.

It can be seen that, whereas DFDs and LDSs present us with a logical and static view of our system, ELHs are considerably more down to earth, and dynamic. The DFD and LDS model the system, while the ELH models the business practices.

13.4 How to derive ELHs

The preceding section is all very well for presenting the theoretical underpinning, but our purpose is to draw, and then use, ELHs. This section describes the procedure for producing the diagrams. As always, I want to emphasize that—like all else in SSADM—this is an iterative process, and it would be less than reasonable to expect you to get this exactly right the first time through.

I shall list the sequence of operations below, and then expand each of them:
- create the Entity/Event Matrix;
- draw first-cut ELHs;
- review ELHs;
- add operations;
- create Effect Correspondence Diagrams;
- add state indicators.

Create the Entity/Event Matrix

As with the drawing of the LDS (Chapter 10), we draw up a grid to provide us with a working document. Along the top of the grid we list all the entities identified during our investigation. The LDS provides us with this information. Operational masters, being access points only, are not counted as entities.

Down the left-hand side of the grid we list all the events that we identify. Our source for this is the set of DFDs already drawn and the associated Function Descriptions.

The lowest level logical DFDs show us the events. We are looking here for those data flows entering the data stores. By backtracking from these data stores, we can identify the event that caused this update.

SS plc gives us the following example. The extract from the DFD shows us the Course Run data store being updated as in Fig. 13.3.

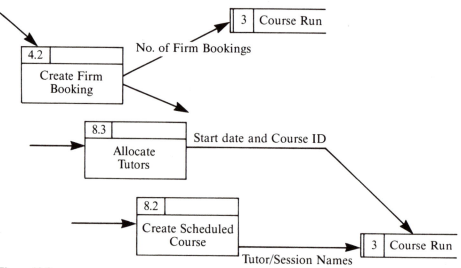

Figure 13.3

Similarly, the input data flow 'Notice of cancellation' is directly responsible for the modification to the data stores 1, 4, 3, 5 as shown in Fig. 13.4.

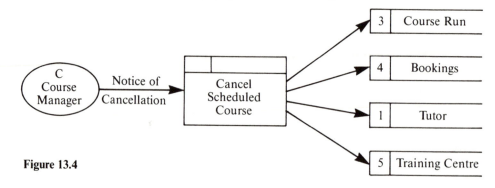

Figure 13.4

Having identified entities and events, we have to bring them together on the matrix. Recognize what effect each event will have on the entities concerned, and whether it is responsible for creating the record, deleting it or changing the value of a field. Choose a simple code to reflect this, such as C–create, M–modify, D–delete, or I–insert, A–amend, R–remove. Mark each relevant intersection on the grid with the appropriate letter. Figure 13.5 shows the small extract from SS plc illustrated in the two previous figures.

Event	Course Run	Bookings	Tutor	Location	Delegate Bookings		
Booking Received	M	C			C		
Course Scheduled	C		M				
Tutor Allocated			M				
Location Booked				M			
Scheduled Course Cancelled	D	D	M	M	D		
Delegate Booking Cancelled	M	M			D		

Figure 13.5

When you have made your first pass at the matrix, review it by checking each dimension. Ensure that each entity is somehow created, somehow deleted, and preferably somehow updated in between. If it is not updated in between, review the purpose of that record. It may be a legitimate life, or you may have misunderstood its purpose in your initial analysis.

Then, ensure that each event does actually impinge on the life of at least one entity. If a row or column on the matrix is left clear, you have made a serious mistake somewhere!

It may well appear, at this stage, that the initial analysis has been faulty, and that processes or data are missing from your earlier DFD or LDS documentation. If this turns out to be the case, put them in and reflect the change in the matrix. We often find that ELHs, as a more detailed tool, expand both DFD and LDS by identifying update anomalies in the earlier stages of analysis.

At the same time, you should review the matrix against the entities and Entity Descriptions—perhaps identifying new attributes for update that were missed earlier. If new attributes—or even new entities—are discovered, are there new knock-on effects on others when the 'new' one is updated?

The analyst will not be sitting in his secluded tower asking himself these questions: each time such an anomaly is found, or query raised, he must return to the User for clarification. At no point is the User regarded as a superfluous agent (or even irritant!) in the process; while SSADM may not be a participative method in the fullest sense, (cf. Mumford's ETHICS (Mumford, 1983)), the constant consultations make it the next best thing.

It can be seen from the above that although the Entity/Event Matrix is a working document only, it takes considerable time and effort to get it right, and only when it *is* right can we proceed to the drawing of the ELH diagrams. To those impatient souls who may ask, is it really worth all the effort and fuss? I can only answer 'yes'! It is *not* worth not taking the effort at this stage: far more time and money will be spent later correcting the mistakes that have arisen from omitting this work.

This exhortation applies to installations who are not using CASE tools. Those using software support should find that inbuilt integrity-checkers will provide the necessary rigour provided by the matrix.

Draw first-cut ELHs

To settle on the 'batting order' for the ELHs, you should have the LDS and the matrix in front of you.

First, using the LDS select all entities that are details only, i.e. that are not themselves masters of other details. These are the first to be done. The last of all will be those entities that are masters only.

As you begin each entity, you will need the Entity Descriptions to recognize which attributes are associated.

Second, from the matrix identify first those events that are responsible for creating and deleting the entity the first and last events in its life. Draw the creation box.

If any of several events may cause creation of an entity, then represent them all as a selection, as in Fig. 13.6.

Using your knowledge of the system, identify possible sequences in which further updates may happen. We know, for example, that in our entity Delegate a Provisional Booking for a course may be made. Later, any Provisional Booking will be changed to Confirmed. Only later are the Joining Instructions sent out.

In this example then, knowledge of the system tells us that the update sequence will be:

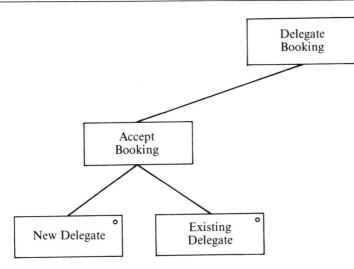

Figure 13.6

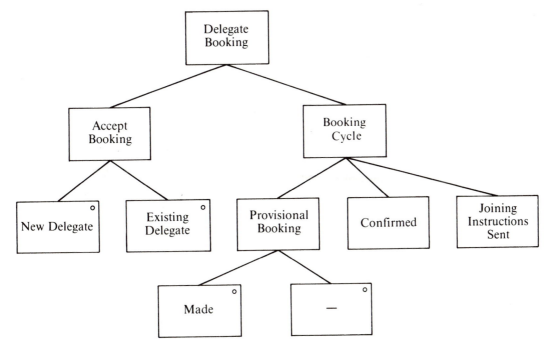

Figure 13.7

1.   make Provisional Booking (possible)
2.   confirm Provisional Booking
3.   send Joining Instructions

as shown in Fig. 13.7.

It is not so straightforward, of course, because of other events, such as possible Delegate Cancellations, Course Cancellations and so on, but the illustration should at least make the principle clear.

Third, using our basic structure of sequence, selection and iteration, draw the simple ELH for your target entity. To help you, although not speed you, you should:

1.   *Look at each entity in turn*   What events are responsible for creating an occurrence of the entity? What events cause an occurrence of the entity to be deleted?
2.   *Look at each attribute in turn*   Identify which events may cause a change to any attribute. Does each attribute have to have a value when the entity is created, or can a value be allocated later? Once a value is given, is it fixed forever, or may it change? If changes are permitted, what event causes them?
3.   *Ask*:
     (a) Is it possible to change relationships with a master entity?
     (b) What events occur to establish a change in an optional relationship?

Once you have established a sequence, confirmed it with your User and drawn the ELH chart to show it, add the deletion events. As with creation, it may be that more than one event may lead to the entity's deletion; if so, examine each possibility to see if a particular sequence of updates or processing must occur, particularly to a related entity, before deletion occurs. Include parallel lives to cater for unpredictable changes to routine detail.

For each entity, you should satisfy yourself that all events that affect it on the matrix have been included.

Fourth, there are occasions when an event might happen out of sequence. That is, an event might occur that causes an earlier part of the sequence to be repeated, or that leaps forward to a future event, missing out several parts of the normal life cycle.

An example might be found in the entity Delegate Booking. It may be that after a confirmation of the booking has been made, the delegate has to change the booking to another run of the course, and therefore has to make a provisional booking. There is more than one effect here: the entity must change sets, from one occurrence of Course Run to another, and the status of Booking changes from confirmed to provisional again. We use a notation known as *quit and resume* to depict this. By the leaf that shows the departure from sequence, we place a sign Qn, and against the leaf where we recommence the life cycle, place Rn (n denotes a corresponding numeric character). Figure 13.8 illustrates the quit and resume for the cited example.

The resume box describes the condition that must be met for the quit to take place. It is important to recognize that quits and resumes are not a lazy way of not bothering about sequencing. If possible, a structure should be built without their use, or at least, with only minimal use. However, in the real world, business rules and practice need to cater for abnormal sequencing of effects through some exceptional event.

On the first pass, though, do not spend time considering all the unlikely exceptions

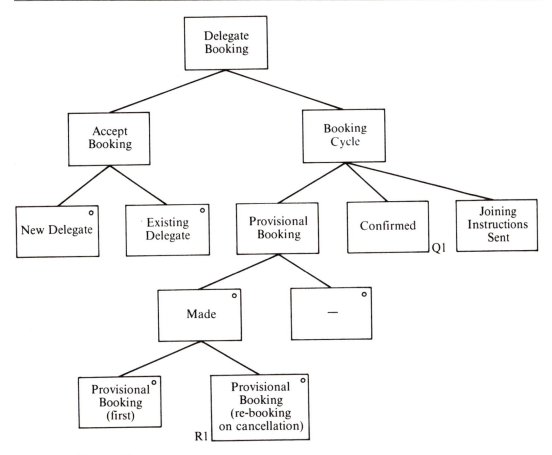

Figure 13.8

and sudden deaths in the life history. That will come with subsequent passes.

It may be that some of the attributes on the entity will not be given a value at the time of creation. Have you identified when they would be updated, or does your chart show them empty still at deletion? If the latter, perhaps there is an event you have missed: back to the User!

Review ELHs

The review of the ELHs should encompass the following points:
- Interactions between entities,
- abnormal death events,
- reversions,
- random events.

INTERACTIONS BETWEEN ENTITIES

Whereas the ELHs were constructed bottom-up, the review is conducted top-down, i.e. beginning at the top of the LDS and following the hierarchy of entities down. This descending review should take account of dependencies between master and detail, particularly the question of disconnection and reconnection between occurrences, and the effect on either of the deletion of the other.

Two structures you should always be aware of can be described as 'n' and 'u' structures, as in Fig. 13.9.

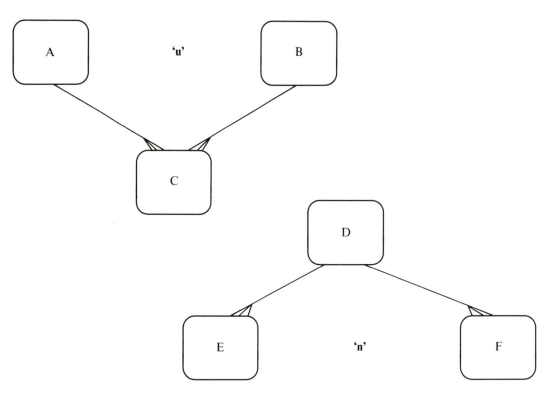

Figure 13.9

ABNORMAL DEATH EVENTS

Not every occurrence of an entity will last its normal life. There are possible exception conditions that can lead to deletion of an entity at any stage in its normal life. Such an exception could be, for example, premature death of an entity, such as a Course Run being cancelled for whatever reason prior to its Start-Date.

This particular exception is catered for by a special use of the quit and resume (see earlier in this section). As this is an event which is unpredictable, and which is not allowed for in the normal life, an obvious and uncluttered notation is called for. This takes the form of a free-standing box at the bottom-right corner of the diagram, with a 'resume *n*' symbol by it. Rather than pepper the diagram with a 'quit *n*' at each leaf, the free-standing box implies that the quit symbol applies universally. To reinforce this, a statement to that effect is placed at the foot of the diagram, as in Fig. 13.10.

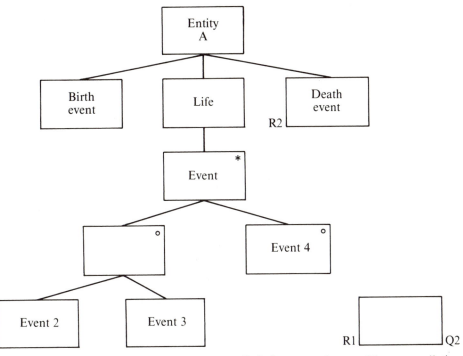

Quit from anywhere to R1 on cancellation

Figure 13.10

On premature death, there are three checks to be made before the entity may be deleted:
1. The death of the master is not allowed until all details are first dead.
2. The death of the master kills all remaining details, or
3. The death of the master has no effect on the detail.
If either of the first two applies, the death of the master must be carried down to all the details.

REVERSIONS
It may be that from a quit point the entity life will begin again, as in the case of a cheque that has been stopped: after the stop, its life may recommence. In that case, there may be a substructure built in, or a new quit and resume from the free-standing box to a selection under the create box, as in Fig. 13.11.

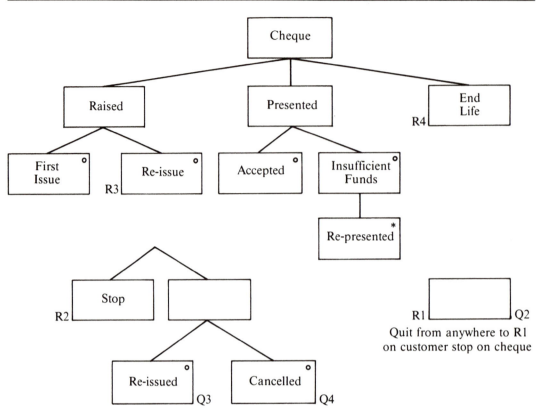

Figure 13.11

RANDOM EVENTS

These are events that affect an entity, but can happen at any time in its life and do not affect the sequencing of the main cycle of that entity. This frequently refers to a housekeeping function in which standing or reference data on an entity is amended. Such data includes addresses, title and so on, which can change without impacting upon the main business of the entity. That is depicted by a *parallel life*. This means that the entity can, as well as having the main sequencing of effects shown, have a separate housekeeping life which records such changes to standing data. Figure 13.12 shows how this is drawn.

Add operations

At each leaf of the ELH, we carry out some operation, whether creating a record, deleting a record or changing the value of a field. The following operations are used in Entity Life Histories. SSADM does not prescribe what should be contained in ⟨expression⟩. It will usually be a form of simple mathematical operation.

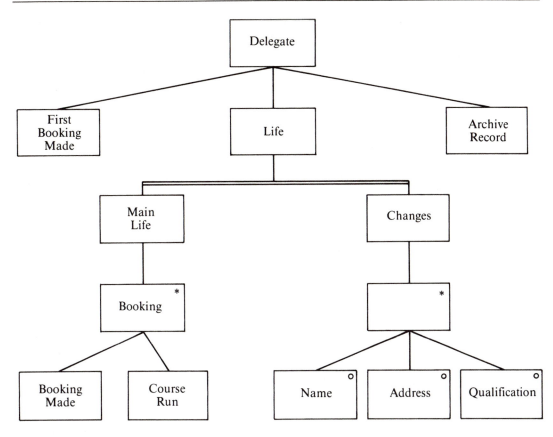

Figure 13.12

Operation	Description
Store ⟨attribute⟩	Set initial value of attribute on creation.
Store keys	Set the value of the primary key, on creation.
Store remaining attributes	Set the values of remaining attributes as input on creation.
Replace ⟨attribute⟩	Change the values of the attribute to the new value input.
Store ⟨attribute⟩ using ⟨expression⟩	Set initial value of an attribute according to a value in the expression.
Replace ⟨attribute⟩ using ⟨expression⟩	Change the value of the attribute according to the value in the expression.
Tie to ⟨entity⟩	Establish a relationship between target entity and a master entity.

Cut from ⟨entity⟩	Remove the relationship between target entity and a master entity.
Gain ⟨entity⟩	Establish a relationship between target entity and a detail entity.
Lose ⟨entity⟩	Remove the relationship beteween this entity and a detail entity.

Any operations to do with validation or navigation do not belong on an ELH. Only include those which cause an update to be made to the entity or its relationships.

The 'Tie', 'Gain', 'Cut' and 'Lose' operations are useful as validation guides: a 'Lose' operation on the master should have a corresponding 'Cut' operation on a detail, and so on.

Number the operations sequentially, and list them underneath the diagram. Then allocate each operation, by its number, to its appropriate place on the ELH. Figure 13.13 shows the ELH for Course Run with its operations listed and allocated.

13.5   Create Effect Correspondence Diagrams

The ELH gives us one view of Entity/Event Modelling: the view of events having an effect on a given entity. Effect Correspondence Diagrams (ECDs) present us with the other view, that of all entities affected by a given event. It is equivalent to the Enquiry Access Paths completed during the LDM, and serves the equivalent purpose for all update processing. In fact, ECDs and Enquiry Access Paths are both compiled during Step 360.

ECDs become the basis for Define Update Processing Model in Stage 5. To describe how to draw an ECD, I shall take an event, Cancel Delegate Booking, and draw the ECD for that, describing the eight steps involved.

First of all, in Fig. 13.14 we see the portion of the LDS affected by this event. Entities affected by the cancellation are: Delegate Booking, Course Run, Transfer Fee, Delegate, Course Title and Branch.

Step 1
Draw a box for each entity affected by the event, as in Fig. 13.15. (If only one entity is affected by an event, the ECD will consist of one box only. The fact that it is a sparse diagram does not invalidate it!)

Step 2
Identify and draw boxes for simultaneous effects. This means that an entity type may be affected in more than one way by an event; one occurrence of the event may affect more than one occurrence of the entity, each one in a different way. If this happens, we say that the entity has different *roles*, each one of which must be named on the appropriate leaf of the ELH.

Our chosen event, Delegate Cancellation, cannot illustrate this particular feature;

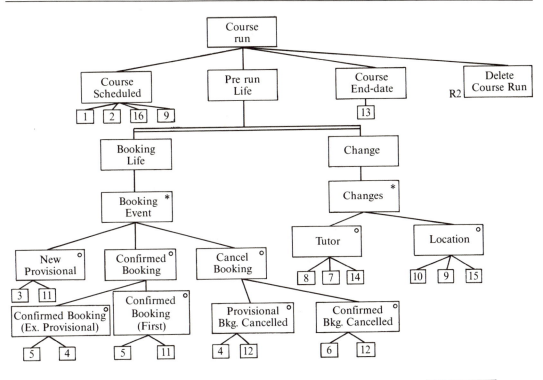

Operations

1. Store Course Run key.
2. Store remaining attributes.
3. Replace No-prov using No-prov + 1.
4. Replace No-prov using No-prov − 1.
5. Replace No-conf using No-conf + 1.
6. Replace No-conf using No-conf − 1.
7. Gain Tutor.
8. Lose Tutor.
9. Gain Location.
10. Lose Location.
11. Gain Booking.
12. Lose Booking.
13. Store summary using input summary.
14. Store Tutor using input Tutor.
15. Store Location using input location.
16. Tie to Course Title.

Quit from anywhere to R1
on course cancellation

Figure 13.13

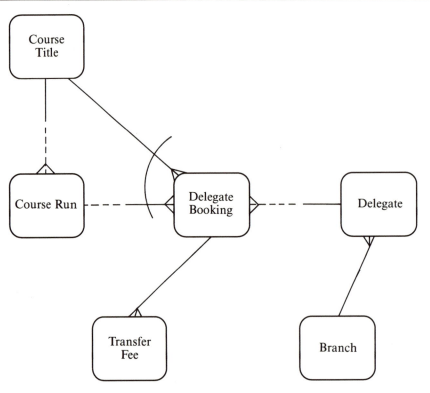

Figure 13.14

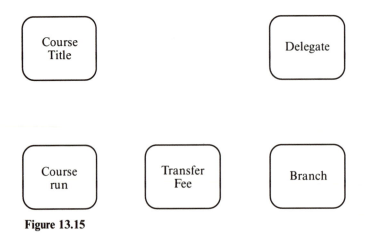

Figure 13.15

unless under the conditions given in Step 4, where some Delegate Bookings may be provisional, and some confirmed.

Step 3

Show optional effects. If an event can affect an entity in more than one way, depending on the condition of the entity, show the possibilities by drawing a Jackson-style selection underneath the entity box.

In our cancellation event, the entity Transfer Fee will be affected differently, depending on whether the cancellation takes place before the penalty date or after. Figure 13.16 shows this.

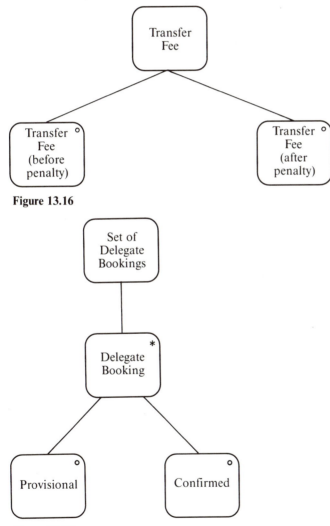

Figure 13.16

Figure 13.17

Step 4

Where more than one occurrence of an entity type needs to be updated on an event (i.e. a number of details of a master), show this as an iteration box. The entity governing the iteration box would be renamed as 'Set of *XX*'. If our Delegate Cancellation came about because the delegate was leaving the company, or the owning unit, all Delegate Bookings belonging to that Delegate would be affected. Figure 13.17 shows this.

Step 5

Identify one-to-one correspondences between effects. This often occurs if an event affects details and master. Ask the question, 'When one occurrence of entity *X* is updated, is only one occurrence of entity *Y* updated?' If so, link the two entity boxes on the ECD with a correspondence arrow. If it is true to say, 'When one occurrence of entity *X* is updated, a set of entity *Y* is also updated', then link entity with Set of *Y* on the ECD with a correspondence arrow.

When Delegate Cancellation is responsible for updating Delegate, it also updates Delegate Booking, or set of Delegate Bookings, Course Run, Transfer Fee and Course Title. Figure 13.18 shows how this affects the diagram.

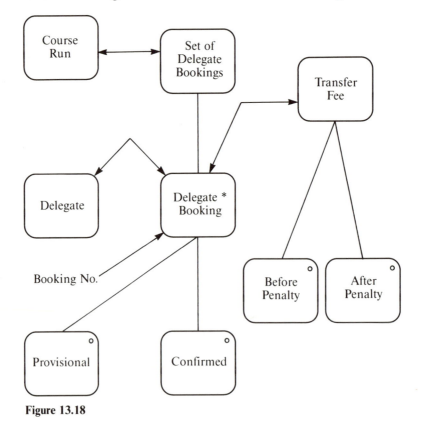

Figure 13.18

Step 6

Merge iterative effects. This means that if an entity is affected iteratively in more than one way by an event, the iterations referring to the same LDS relationship, the effects should be merged into a single structure. There is no such example in our Delegate Cancellation event.

Step 7

Include non-update entities. It may be that to carry out all the effects, some entities must be accessed for navigational purposes only. If the ECD cannot currently access all the data needed to reflect the effects of the given event, then add the other entities to allow this.

Step 8

Add event data. Show the input data to the update process, against the entity which is the entry point to the LDS. The data may be the key of the entity, or the key plus several attributes. It may even be non-key attributes, although these tend to trigger enquiries rather than events.

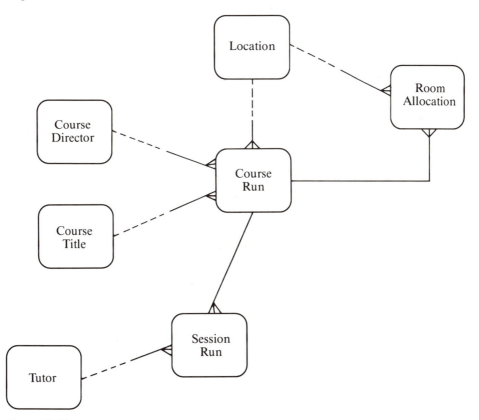

Figure 13.19

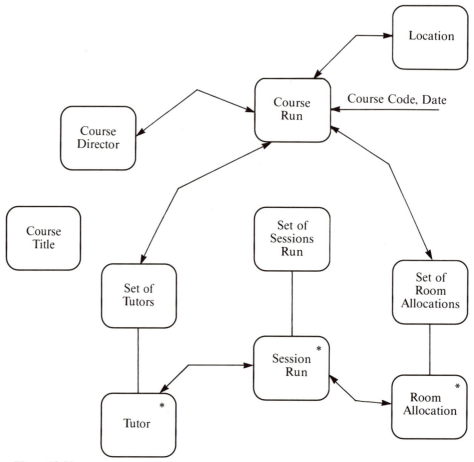

Figure 13.20

Let us take the event Insert Course Run. Figure 13.19 shows the relevant part of the Logical Data Model, and Fig. 13.20 shows the completed Effect Correspondence Diagram.

13.6 Add state indicators

What are state indicators?
Expressed most simply, a state indicator is a single-byte attribute that is changed each time the entity is updated. It takes the form of a numeric digit, and is written under each *leaf* of the ELH to reflect the values. Each time an event affects the entity, the value of the state indicator is changed to indicate the current state of the records. The following example will make this more clear.

The rules for adding state indicators act as explanation for Fig. 13.21.

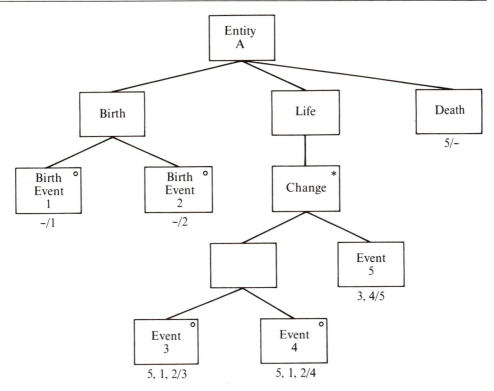

Figure 13.21

1. Each leaf has two values, a 'before' and 'after' value. The value is conventionally numeric, starting from 1 at creation, and increasing for each new event that changes it.
2. The first and last values, i.e. on creation and deletion, are 'null' values.
3. On an iteration, the valid previous indicators include the new valid indicator, as in the iteration in Fig. 13.21.
4. If a value is not changed, as in a parallel life event, or if any previous value is valid, this is shown as an asterisk.
5. When using a quit and resume, the state indicator is set *before* the quit, so the resume box shows that new value as a valid previous value.

State indicators are added to the diagrams in Step 520, and are the last thing to be done to the ELH.

Why state indicators?
The value of state indicators is twofold: as they are actual fields on the record, they can be used for validation purposes and as a secondary index.

Before an update is made on a particular occurrence, the state indicator can be interrogated to be sure that the record is in an acceptable state for that particular amendment. This can save considerable effort in the validation stage of a program.

As each indicator value means that the record is at a particular point in its life, so it can be easier to retrieve all records at a certain stage of processing by using that byte as the search key, rather than by formulating a complicated, multi-field search strategy. Obviously, that is not invariably the case: the state indicator value will tell us that delegate John Matthews has confirmed a booking for a particular course, but not how many provisional places there are still on that course.

SUMMARY

Entity/Event Modelling gives us our third view of the system, supplementing the functional view of the DFM and the data view of the LDM.

It provides us with a dual viewpoint of the effects of the events that drive our system. The Entity Life History shows the entity viewpoint of the sequence of events that affect it; Effect Correspondence Diagrams show the event viewpoint of the entities that are affected.

The outputs from this activity are input to the Logical Design activity.

EXERCISES

1. The environment below gives more detailed information on the college library described at the end of Chapter 10.

   An entity Loan is created when a Member borrows a book. At any time before loan expiry, the book may be renewed up to a maximum of three times.

   If anyone has reserved the book, the Loan cannot be renewed.

   If a book has not been returned within one week of Loan expiry the first of three reminders is sent.

   If the book has not been returned within two weeks of the third reminder, the Member is put on a blacklist.

   As soon as the borrower is blacklisted, the Loan entity is deleted and an Outstanding Loan entity created in its place. Otherwise the Loan entity will be deleted six months after the book has been returned.

   The blacklisting will end if the book is returned, or the price of the book is paid. If neither happens within one year of the blacklisting the Member entity is deleted and a Banned record created instead, with the Member's identifier for key.

   Create an Entity/Event Matrix based on this outline. When the matrix is complete, draw the Entity Life History for the entity Loan.

   Attributes for Loan are:
   - Book Identifier/Member Identifier
   - Date Loan Created } key
   - Renewal Indicator (values 0, 1, 2, 3)
   - Date Renewal Expires
   - Reminder Indicator (values 0, 1, 2, 3)

- Reminder Date
- Title Reserved Indicator

2. Using the above information, draw an Effect Correspondence Diagram for the event Make Loan.

14. Dialogue Design

14.1 Aims of chapter

In this chapter you will learn:
- How and when to identify key dialogues.
- How and when to identify and record the different User Roles.
- The notation for I/O Structures.
- The notation for Dialogue Structures.
- How to identify and record Logical Grouping of Dialogue Elements.
- How to identify navigation paths between Logical Groupings of Dialogue Elements.
- How to design Menu and Command Structures.

14.2 Where Dialogue Design is used in SSADM

The different elements for this technique are gathered and identified right from the beginning of the analysis, in Step 110.

Step 110—Establish analysis framework In this step we create the User Catalogue.

Step 120—Investigate and define requirements The User Catalogue is amended here, in the light of the further investigation.

Step 310—Define required system processing Yet again, we update the User Catalogue according to our findings, and the establishments of user requirements.

Step 330—Derive system functions At this step, we identify the required dialogues in the new system. Activities include identifying the functions associated with various User Roles, and specifying the I/O interfaces associated with each function.

Having identified the dialogues for the new system, we go on to identify those regarded as critical. This activity closes the identification of dialogues. The actual design takes place in Stage 5.

Step 420—Select Technical System Options In this step we create an Application Style Guide, if no such guide exists locally. There is little here in the way of Dialogue Design, but the guide will be used in its creation in Step 510.

Step 510—Design User Dialogues   This is the step when the dialogues are created. The activities involved here are:

- Create Dialogue Control Table.
- Create Dialogue Element Descriptions.
- Create Dialogue Structures.
- Create Menu Structures.
- Create Command Structures.

All of the above activities will be expanded in this chapter.

The pattern of activities is shown in Fig. 14.1.

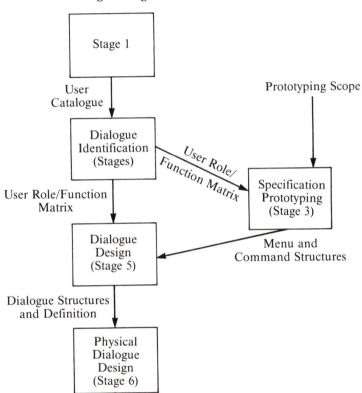

Dialogue Design

Stage 1

User Catalogue

Dialogue Identification (Stages)

User Role/Function Matrix

Prototyping Scope

User Role/ Function Matrix

Specification Prototyping (Stage 3)

Dialogue Design (Stage 5)

Menu and Command Structures

Dialogue Structures and Definition

Physical Dialogue Design (Stage 6)

Figure 14.1

Use in SSADM

Although the chapter is entitled 'Dialogue Design', the actual design is the last activity to be carried out. Under this umbrella heading come the activities of Dialogue Identification and Dialogue Specification as well as Dialogue Design.

The specification of dialogues is closely tied in with User Roles, and the identification of these roles is an important part of this exercise.

Inputs to Dialogue Design are as follows:
- Function Definitions
- Installation Style Guide
- I/O Structures
- Menu and Command Structures
- Requirements Catalogue
- User Catalogue
- User Role/Function Matrix
- User Roles

Outputs from Dialogue Design are as follows:
- Application Style Guide
- Command Structures
- Dialogue Control Table
- Dialogue Element Descriptions
- Dialogue-level help
- Dialogue Structures
- Menu Structures
- Requirements Catalogue
- User Catalogue
- User Role/Function Matrix
- User Roles

Relationships with other techniques

Figure 14.2 illustrates the relationships between Dialogue Design and other SSADM techniques. The two closest relationships are Function Definition, which drives the dialogues, and Specification Prototyping, which uses the initial identification of dialogues as its input, and whose own output is fed into Dialogue Design.

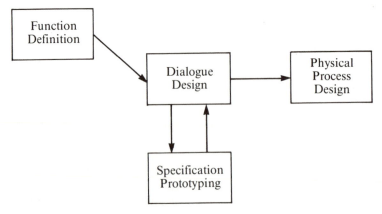

Figure 14.2

User Catalogue

Project/System SS. plc	Author	Date	Version	Status	Page of

Job Title	Job Activities Description
Course Manager	To be responsible for contents of Course, and in liaison with Course Scheduling, to arrange for the dates of courses to be run. To authorize cancellation of a course.
Bookings Manager	Responsible for three Bookings Clerks. Team receive Bookings Requests from Nominating Managers to place a number of Delegates on a Course Run, or on a Waiting List for a Course Title.
Bookings Clerk	Receives Bookings request. Places Delegates on Waiting List or Course Run. Updates Wallchart regularly and adhoc, with Place Numbers. Cancels Delegate Booking. Sends out Joining Instructions
Course Scheduling Manager	Responsible for two Clerks. With Course Managers, arranges Schedule of Courses for three months ahead. Allocates Tutors for each Course and Session.

Figure 14.3

14.3   Procedures

Dialogue identification

PRODUCE USER CATALOGUE

This document is established at the beginning of the analysis, in Step 110. It describes the users of the planned system—and, if applicable, the current system—by job title and task. Figure 14.3 illustrates the User Catalogue.

This catalogue is next used to identify the User Roles in the new system, and so it concentrates on two facts:
1.   the Users in the target population, and
2.   the tasks and functions performed by each User.
Note: we do not identify the users as individuals on the catalogue but as actors of a function. They are identified by job title.

IDENTIFY USER ROLES

By 'User Role' we mean a collection of job holders who carry out tasks in common. The User Roles are identified from the User Catalogue: two pointers that will help you to identify User Roles are:
1.   users with similar job descriptions, and
2.   users who communicate with the same external entities.
Examine the User Catalogue with the above indicators in mind to identify all the User Roles. Some overlaps will be obvious, others only clear after analysis or further investigation. Examples in SS plc would be as shown in Fig. 14.4.

IDENTIFY THE DIALOGUES REQUIRED

Using the User Catalogue, the User Roles and the Function Definitions, draw up a User Role/Function Matrix. An example of this diagram is shown in Fig. 14.5, which develops the SS plc User Roles above.

Each intersection on the matrix represents a dialogue, and so each row identifies all dialogues a given user requires. Check the matrix with the Users, to make sure that you have indeed identified all the dialogues each wants, and that none are superfluous.

It is quite likely that some dialogues will be optimized, i.e. if different User Roles are performing the same function, and use the same data, you may develop just one dialogue for all.

IDENTIFY CRITICAL DIALOGUES

Some dialogues may be considered critical to the success of the system, for various reasons. Go back to the matrix, and circle those intersections you identify as critical. Use the following criteria to help you identify a critical dialogue:
- Dialogues that users consider critical for their work.
- Dialogues that are shared between many user roles.

User Roles

Project/System SS. plc	Author	Date	Version	Status	Page of

User Role	Job Title	Activities
Booking Clerk	Booking Clerk (customer) Booking Clerk (Lecturer)	Make provisional/confirmed bookings for customers or lecturers on courses. Make amendments or delete customer, resource or machine details. Deals with enquiries.
Booking Management	Booking Clerk (management) Bookings Manager	Produce reports. Book, amend, delete any course, resource or machine. Maintain course and resource details.
Invoicing	Booking Clerk (sales) Invoicing staff	Price and post invoices. Maintain price details. Produce statistical reports.
etc.	etc.	etc.

Figure 14.4

User Role/Function Matrix

User Roles \ Functions	Insert New Booking	Delete Cancelled Bookings	Book Rooms For Courses	Replace Cancelled Delegates	Reschedule Delegate Booking	Allocate Tutor	Create New Course	
Course Mgr							✕	
Bookings Office	✕	✕		✕	✕			
Scheduling Office			✕			✕		

Figure 14.5

- Dialogues that read from or write to many entities.
- Dialogues that access or input large amounts of data.
- Dialogues that will be used frequently.
- Dialogues that are central to the key business functions.

With this identification of critical dialogues, the dialogue identification activities are complete.

Dialogue Design

IDENTIFY DATA ITEMS

Our input to the design activities is the set of I/O Structures built up from the DFD set and Function Description. Each I/O Structure becomes a Dialogue Structure. Figure 14.6 illustrates this for the dialogue Make Booking.

Next, identify the data items for each Dialogue Structure. This information will be recorded already on the documents supporting the I/O Structure. List each of the items on SSADM standard form.

IDENTIFY LOGICAL GROUPING OF DIALOGUE ELEMENTS (LGDE)

We have to navigate between elements of our dialogues. To aid the design of these navigation paths, look for a logical grouping of the elements in the structure. Do this in consultation with the User. As there is no hard-and-fast algorithm to achieve this, the following guidelines may help you:

- Dialogue elements that occur in a sequence may be grouped together as a LGDE.
- Dialogue elements that are not adjacent/sequential on the structure should not be grouped into one LGDE, unless perhaps the data elements between are included.
- A LGDE must *not* contain only a subsection of a dialogue element.

Generally, a LGDE embraces both input and corresponding output of a dialogue element. If an input involves a considerable amount of data, then there is a strong case for making the LGDE embrace the input only.

When you have identified the LGDEs and annotated the structure, give each one an ID, numbering sequentially from the left. Figure 14.7 shows the Dialogue Structure for Make Booking with LGDEs identified and annotated.

IDENTIFY NAVIGATION BETWEEN LGDES

In our Make Bookings example, we have identified four LGDEs. We need to navigate our way between them to complete the dialogue. If the dialogue structure consists of a simple, uncomplicated sequence, navigation is no problem. If there are null selections and iterations, however, identifying and specifying valid paths from one element to the next is more of a problem.

Figure 14.8 shows this for another dialogue: Delete Delegate. This has five LGDEs, and contains iterations and null selections. Wherever there is an iteration that may have null occurrences or a selection, we recognize the navigation problem.

The SSADM document to achieve this is the Dialogue Control Table (Fig. 14.9). This illustrates the possible navigation paths for our Delete Delegate dialogue. You will see on the table a column that classifies LGDEs as *mandatory* or *optional*. If the elements succeed one another without possible variation, i.e. a simple sequence, the LGDE is mandatory. If the element contains a null selection, or an iteration which may have null occurrences, it is optional.

One last piece of documentation that requires to be completed is a *Dialogue Element Description*. This identifies all the data items in each data element, and

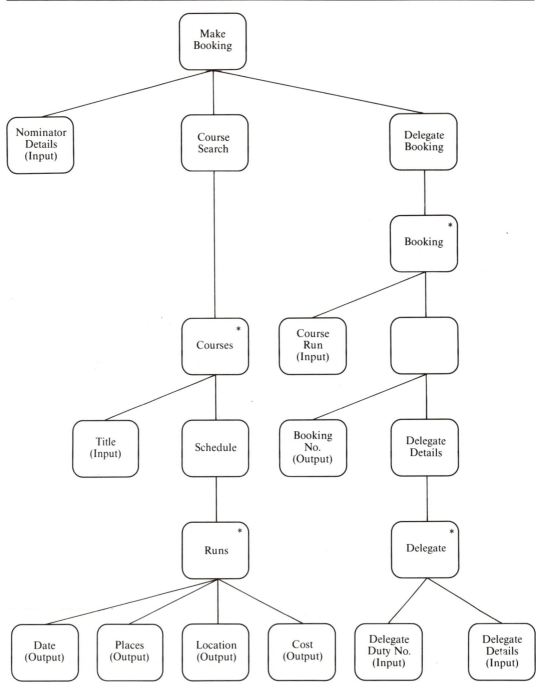

Figure 14.6

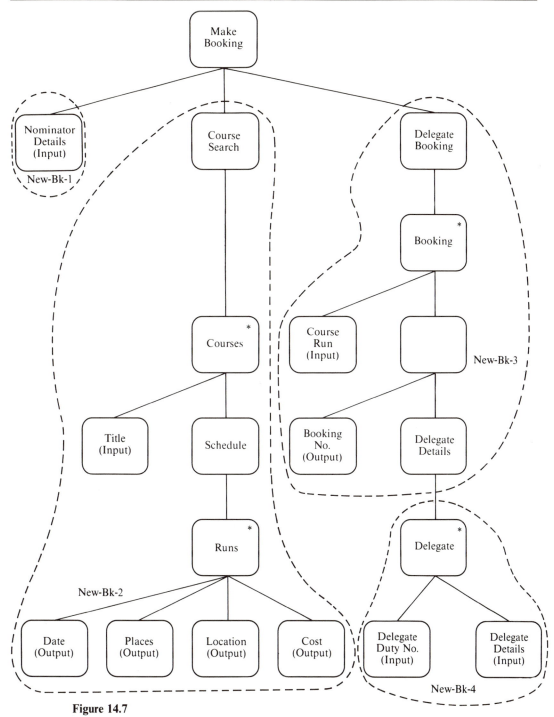

Figure 14.7

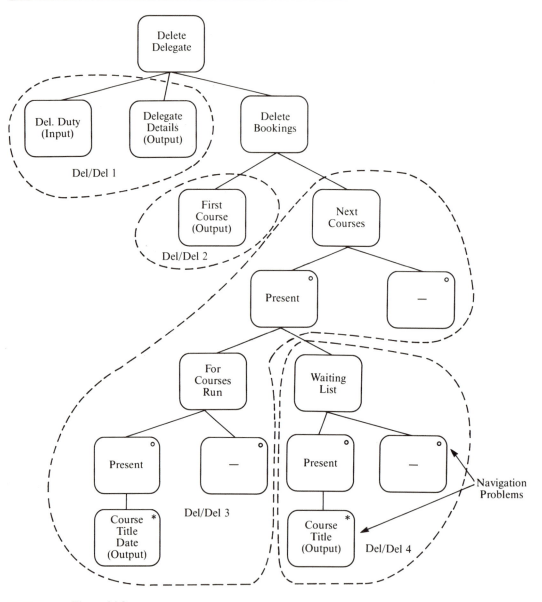

Figure 14.8

Dialogue Control Table

Dialogue Names:	Delete Delegate	User Role:	Bookings Officer

LGDE ID	Occurrences			Default Pathway	Alternative pathways		
	Min.	Max.	Ave.		Alt.1	Alt.2	Alt.3
Del/Del 1	1	1	1	X	X	X	X
Del/Del 2	1	1	1	X	X	X	X
Del/Del 3	0	10	2	X	X		
Del/Del 4	0	10	3	X		X	
% Path usage				70	5	5	20

Figure 14.9

classifies each element as mandatory or optional. Figure 14.10 illustrates the Dialogue Element Description form for the first parts of Delete Delegate.

DESIGN MENU AND COMMAND STRUCTURES

Menus

Menus are a device, usually hierarchical, which give a user or user group access to permitted online applications. This design activity is to group those applications that logically fit together, either by virtue of belonging to the same function, or by being legally available to the same user group.

The SSADM tool we use for constructing menus is the User Role/Function Matrix. Figure 14.11 repeats the matrix seen earlier in this chapter. The first thing to do is to identify all of the functions performed by a User Role.

Having identified all the possible dialogues for that User Role, we build them into a hierarchy by grouping them together in logical order. The actual dialogues identified will form the lowest levels of the hierarchy. As with the LGDEs, there is no fixed algorithm or calculation for deciding on the groupings, so the following rules of thumb will help:

- Bring together in one group those dialogues which logically go together. Refer to the DFD for the Required System to help with this; the groupings of the processes on that diagram will give you pointers, as will consultation with the Users, of course.
- Reflect the User's way of performing the tasks. Our structure should support whatever sequence the User chooses to carry out the task. If this means invoking the one dialogue from different places in the menu hierarchy then that must be permitted and shown.
- Each grouping in the hierarchy may lead *either* to another lower level menu *or* to a dialogue on the bottom level.
- There is no need for groupings to have the same number of items: logic, not symmetry is our aim.

Another important input to this activity is the result of any prototyping exercise that involved the use or design of menus.

Having built up our menu hierarchy, we represent it as a tree structure, with the User's entry point to the system the single node at the top. See Fig. 14.12.

We can see from this diagram, that each node may lead either to further menus, or directly into the dialogue. The thing it may *not* do is lead from a dialogue to a lower level menu.

Each menu node on the diagram is depicted by a rectangular box, referenced MEN *nn*, where *nn* is a number. Dialogues are depicted by a round-cornered box, referenced DIAL *nn*, where *nn* is a number.

At the end of the structure for a given User Role, cross-check with the User Role/Function Matrix. Every cross on the User Role's row must be represented by a dialogue box at the bottom of one branch of the hierarchy.

Every User Role on the matrix must have one Menu Structure.

Dialogue Element Descriptions

Dialogue Name — Delete Delegate

User Role — Bookings Officer

Dialogue element	Data item	Logical grouping of dialogue elements ID	Mandatory / optional L G D E
Del. Duty	Delegate Duty Code	Del/Del 1	M
Delegate Details	Delegate Name Delegate Address Delegate Branch		
First Course	Course Title Start Date	Del/Del 2	M
Course Run	Course Title Start Date	Del/Del 3	O

Figure 14.10

User Role/Function Matrix								

Functions / User Roles	Insert New Booking	Delete Cancelled Bookings	Book Rooms For Courses	Replace Cancelled Delegates	Reschedule Delegate Booking	Allocate Tutor	Reschedule Course Run	Confirm Booking
Bookings Officer	X	X		X	X			X
Scheduling Officer			X			X	X	

Figure 14.11

Command Structures

Command Structures show the directions that control can take when a dialogue is completed. This means that dialogues can be invoked with or without menus.

The Command Structure is essentially a form—one for every dialogue—that documents all the possible directions a User can take on completing that dialogue. Figure 14.13 shows the Command Structure for Cancel Delegate Booking.

At the close of the dialogue the user has three courses of action open:

1. To return to the main user menu.
2. To continue the same dialogue with another transaction; this means a return to the beginning of the dialogue, but not back to the menu.
3. To quit directly to a related function such as Amend Transfer Fee, or Cancel Delegate.

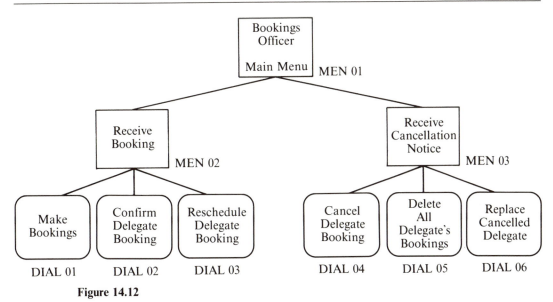

Figure 14.12

The Command Structure listing these must contain only items valid for that particular User Role. Refer to the User Role/Function Matrix for confirmation that you have offered all the choices, and only the valid choices.

The actual mechanism for navigating through the dialogues will be left to the Physical Design. Such mechanisms will include the use of function/control keys or typed commands to move the User from one part of the menu/dialogue to another.

When making decisions about the Command Structures, take into account such factors as:

● constraints imposed by selected hardware;
● relative frequency of dialogues, and dependencies between dialogues.

DEFINE DIALOGUE-LEVEL HELP

At this point it may be useful to define the navigation help procedures associated with each dialogue. This is not the place to design help screens; that comes with Physical Design, and also prototyping activities. Rather it is to identify the requirements for help facilities that the User may wish.

Help may be wanted on three levels:

1. *Context*   Where am I in the dialogue/menu structure?
2. *Job-related*   What do I do now?
3. *Navigational*   Where's the way out?

If these facilities impact on the associated Dialogue Structures, then the structures must be altered and the supporting documentation revised.

Function: Cancel Delegate Booking.	User Role: Bookings

Option	Dial/Menu	Dial/Menu name
Delete All Bookings	Dialogue	Delete All Delegate Bookings
Reschedule Delegate	Dialogue	Reschedule Delegate Booking
Find Replacement	Dialogue	Replace Cancelled Delegate
Quit to Menu	Menu	Bookings Officer
Cancel Another Delegate	Menu	Cancellation Notice

Figure 14.13

SUMMARY
Dialogue Design is an activity that takes place towards the latter part of the analysis-and-design cycle, yet preparation begins at the very start in Step 110.

SSADM is concerned with improving the specification and design of online systems. The emphasis now placed on the User Roles, and Function Definition enables us to carry out this specification more rigorously than before. We can develop the dialogue structures from what has been uncovered before, rather than begin a new area of analysis, with the attendant risks of missing important requirements previously identified.

15. Specification Prototyping

15.1 Aims of chapter

In this chapter, you will learn:
- The place of Specification Prototyping in SSADM.
- The reasons for carrying out Specification Prototyping.
- The inputs to Specification Prototyping.
- The products of Specification Prototyping.
- The conduct of Specification Prototyping sessions.

15.2 Where Specification Prototyping is used in SSADM

Step 350—Develop Specification Prototyping

Use in SSADM

In SSADM, Specification Prototyping has its own prescribed place and procedures. There are other forms of prototyping that can be used in systems development, whether in a SSADM project or not. The other approaches are not described here. Guidelines on the use of these techniques may be found in the Prototyping Interface Guide referred to in Chapter 2.

Specification Prototyping is a method of examining the workings of the Requirements Specification with the User, with the intention of trapping errors, and helping the User to identify new requirements. It is not a method of developing or designing a physical system.

Critical dialogues are the first to be selected for the prototyping process. After, any other dialogues that the User wishes to see trialled can be tested.

Each dialogue chosen is run in a Specification Prototyping session, i.e. a computer simulation of the dialogue is placed, using dummy data and inputs. Any changes to the dialogue that surface as a result of this are made, and an amended version played. All such changes are documented, as is the Requirements Catalogue.

Inputs to Specification Prototyping are as follows:
- Data Catalogue
- I/O Structures
- Installation Style Guide
- Prototyping scope
- Requirements Catalogue
- Required System LDM
- User Role/Function Matrix

Outputs from Specification Prototyping are as follows:
- Command Structures
- Menu Structures
- Prototyping Report
- Requirements Catalogue
- Prototype Demonstration Objective Documents
- Prototype Pathways
- Prototype Result Logs
- Screen Formats

Figure 15.1 illustrates the relationship between Specification Prototyping and other SSADM techniques.

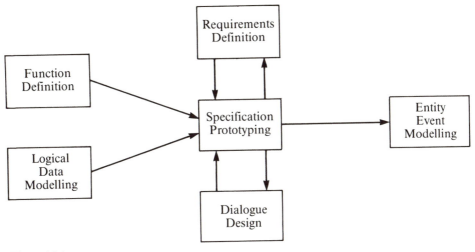

Figure 15.1

15.3   Preparation for the prototyping session

Selection of the tool

As the technical environment may not be known at this point, the target implementation environment is obviously not a candidate for use in the prototyping session. Plenty of tools exist on the market, sharing several essential features: screen painter, data dictionary and online navigation.

Input to the prototyping activities will include some SSADM products such as the LDM for the required system. If you are using a CASE tool for the project development, that tool itself may be used for the prototyping; at least, it should be able to interface with the prototyping tool.

Whatever tool is used, it should be chosen and procured early in the project life. If the IS strategy has dictated the technical environment before TSO, then the prototyping tool should be chosen to simulate that environment as closely as possible. If no indication has been made as to environment, then obviously that cannot be done.

Scoping the prototyping activities

This activity should be carried out at the start of the whole project. The first thing to do is to identify any need to perform Specification Prototyping. If a project has one of the following characteristics, then prototyping will probably not be appropriate:

- The new system is being translated onto the current system directly. If the change is merely an upgrade of hardware/software, there will be no need to trial the Requirements Specification this way.
- The nature or size of the project does not justify the cost of resources for Specification Prototyping.

If prototyping is justified, is it screen prototyping, or report output prototyping? Probably both will be required for the system. In any case, you should establish the following before you begin:

SCREEN PROTOTYPING

- What is the likely level of online activity? If there is considerable interaction, Prototyping will help validate the Specification requirements.
- What is the likely level of data manipulation on screen? If for any one function the activity is large, Specification Prototyping will assist in validating and collecting the User's needs.
- If the online interaction is poorly thought out would that have a detrimental effect on the business, or would it just be inconvenient at a local and trivial level? If the former is the case, then prototyping would be valuable.

REPORT OUTPUT PROTOTYPING

- If an output from our target system is to serve as input to another, prototyping can help to ensure that the output meets the requirements.
- If an output has to meet certain statutory requirements, tax return forms for example, again prototyping can help validate its content and format.
- If the requirements for the report have been described in vague terms, for whatever reasons, prototyping will help define levels of accuracy, optimum format, etc. as the User sees the possible versions of the report and tries to use it.

Setting up the team

The team to carry out the Specification Prototyping must be defined well before the activity is due to start, so that the management structures can be put into place in good time.

The team should comprise a team leader and two other analysts, who between them will serve the roles of implementing the prototyped model and demonstrating it to the User. There is no absolute need for two analysts, of course; one would be sufficient if the project were not too large; a second team member, however, does provide an objective view of a prototype designed by someone else. The effect of this would be that during demonstration the analyst would be more sensitive to the User's requirements, and less defensive about the product.

The team leader's responsibility, apart from standard supervisory duties, should be for the following:

- approving the choice of dialogues and reports to be prototyped;
- agreeing feedback from the demonstration sessions;
- deciding when to close a prototype cycle for a product;
- notifying changes to SSADM documentation, as a result of a prototyping cycle, to the relevant authorities;
- making the final report to management, summarizing the outcome of the Prototyping cycles and reasons for decisions made.

15.4   Procedures

This section describes the procedures followed in Step 350, and the products of the prototyping exercise.

Define the scope of prototyping

Management will normally have specified the areas, specific dialogues and output reports that are to be prototyped. This document is not the final word, however. The team's first task is to define the scope of the activity, using the outputs from Step 330:

- I/O Structure for each function,
- User Role/Function Matrix, defining criticial dialogues.

The critical dialogues should be studied to see if they are appropriate for Specification Prototyping. Confirm the decisions with the Users, and ask for further dialogues that they want to see prototyped. The only constraint on the team's agreement should be budget and timescales.

Some reports should also be prototyped, particularly those using formats dictated by external bodies, such as the Inland Revenue for their PAYE forms, or the Bank Automatic Clearing System (BACS) for their standard forms.

Produce the initial prototypes

The tool for this activity is the Prototype Pathway. This is drawn up for each User Role, and takes that role through the pathway for a particular dialogue, from initial menu, or command, to completion of the function task. Its main purpose is to show the linking of all screens and reports in the prototype.

Figure 15.2 demonstrates a Prototype Pathway for the dialogue Cancel Delegate Booking. This example shows the path the User will take, and identifies each component as a menu, screen or report. Each component will have its own unique identifier and Function Description.

The components must be identified first, from the Logical Grouping of Dialogue Elements (LGDE), constructed during Dialogue Definition, and from the Requirements Catalogue. Once the screens and reports are identified, the pathway is completed by putting them in a logical sequence, and linking them with a menu or command structure.

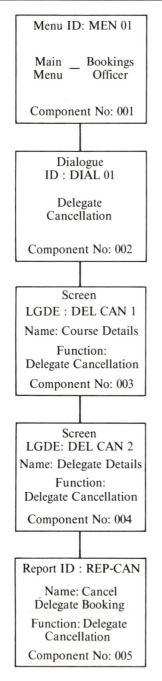

Menu ID: MEN 01

Main _ Bookings
Menu   Officer

Component No: 001

Dialogue
ID : DIAL 01

Delegate
Cancellation

Component No: 002

Screen
LGDE : DEL CAN 1

Name: Course Details

Function:
Delegate Cancellation

Component No: 003

Screen
LGDE: DEL CAN 2

Name: Delegate Details

Function:
Delegate Cancellation

Component No: 004

Report ID : REP-CAN

Name: Cancel
Delegate Booking

Function: Delegate
Cancellation

Component No: 005

Figure 15.2

Once the Prototype Pathways have been identified, they can be implemented on the prototyping tool. The menus may have been implemented already, either by the team in anticipation of this step, or they may be in place in template form, as a feature of the particular tool used.

Once we reach this point, we must begin to think in terms of screen design and ergonomics. Screen design itself is a discipline outside SSADM, and therefore beyond the scope of this book. Local Installation Style Guides, which should be produced to support this activity, will give guidance to the designer, ensuring that the screens and pathways in these prototypes adhere to local standards.

The designer's responsibility here is to enable the User to concentrate on the content of the screen, rather than be distracted by layout.

During the prototyping sessions, the output data items should be validated, usually against the LDM and Data Item Descriptions. Some data items will be derived from, for example, calculations: the formulae should be recorded with the prototyping documentation, and also on Elementary Process Descriptions and/or Function Definitions.

Prepare for the prototype demonstration

A Prototype Demonstration Objective Document should be completed for each pathway identified. Figure 15.3 gives an example for a Prototype Demonstration Objective, for the pathway illustrated in Fig. 15.2, Cancel Delegate Booking.

The purpose of the document is to force designer and user to understand the objectives of each prototype before the demonstration, rather than play about with the prototype because objectives have not been defined. This will lead to a more fruitful demonstration session.

For each component on the pathway, its assumptions, purpose and queries are listed. The points for checking and discussing during the session are all listed.

Test data and inputs for each of the prototype dialogues must be prepared, to show that the inputs and outputs are as required, and that the processing is accurate.

False data must also be prepared to show how the dialogues can handle erroneous input.

The prototypes should be tested against the User requirements, to make sure that the functional scope, the inputs and outputs are all accurate.

It must be remembered during this exercise that the prototype is not the front end of an actual system, but a simulation of the interface only.

Demonstrate and review prototypes

The User Roles involved in each dialogue will have been identified during the early preparatory work on Dialogue Design. Representatives from the relevant User Role should now attend the prototype demonstration of their pathway, with preferably two members of the prototyping team.

The Prototype Demonstration Objective Document will be used by designer and User as a checklist of all discussion points for each component.

A Prototype Result Log will be created to record the results of each demonstration.

Prototype Demonstration Objective Document

Document No: 021	Prototype Pathway No: 0001
Function name: Booking Cancellation	User Role: Bookings

Agenda

1. User has not been involved in prototype demonstrations before, so:

(a) Discuss area covered by the prototype.
(b) Explain procedure of prototyping.

2. Arrange for redemonstration.

3. Clarify navigation details of components 3 and 4.

Component No.	Component queries
0001	Check OK
002	Is correct screen returned?
003	Are all details on the screen?
003	Is the display intelligible?
004	Are the delegate details all displayed?
005	Does the report need to show new number of places free?

Figure 15.3

The log will record each User request, for each screen, report or menu, for each version of the prototype.

The results of each component demonstration will be recorded during the demonstration session, and a note made of the kind of change needed to satisfy the requirements.

Every result on the log will be annotated with a code that classifies the change required. There are seven codes that can be applied here.

The codes that classify changes are:

N   No change required.
C   Cosmetic changes only. These affect the presentation, not the content of the components. Time should not be spent making cosmetic changes only: wait until more substantial changes are to be made, and incorporate the cosmetic alterations then.
D   Changes that affect the dialogue only.
P   Changes that affect the Prototype Pathway. It may be that the changes will affect other documents, such as I/O Structures for the relevant function.
S   The results highlight a weakness in the standards: perhaps they should be investigated with a view to having them altered.
A   Results have indicated that the analysis may be faulty. This is a serious error that will cause a suspension of Prototyping activities while the matter is referred to management. At worst it may cause a second pass through some or all analysis stages.
G   A change is indicated which has ramifications beyond the application. It is possible that they would affect the organization's working practices. If that is the case, then again the matter must be referred to management before action is taken.

Figure 15.4 shows the Prototype Result Log for the prototype demonstration of Cancel Delegate Booking.

On completion of the logs, the team leader must make a decision on further action. There are three questions on which to base the decision:
1.   Will further demonstrations be needed, or will no purpose be served by continuing?
2.   Should the timescales be extended, or more resources allocated? If the answer to these is 'yes', project management must be approached to sanction the decision.
3.   Have any problems surfaced that must be drawn to management's attention?
If the Specification Prototyping is to continue, the team leader will instruct the demonstrators to make the necessary changes and arrange the new demonstrations.

Perform updates to supporting SSADM documentation

Any changes to SSADM documentation that result from the prototyping exercises should be implemented by the team leader. Such changes may be identified on each pass at a prototype. The products to change may be I/O Structures, Required System LDM, Requirements Catalogue, DFM, Enquiry Access Paths, Effect Correspondence

Prototype Result Log

Prototype Result Log No. 021		Prototype Pathway No. 001	
Function name: Booking Cancellation		User Role: Bookings	

Component No.	Result No.	Result Description	Change Grade
001	01	Satisfactory	N
002	01	Satisfactory	N
003	01	Course date should be shown	D
003	02	Screen cluttered in the centre	C
004	01	Too many delegates shown	A

Figure 15.4

Diagrams, or even User Role/Function Matrix. Suggested changes to such documents should be reported and investigated promptly, to see whether they will in fact give the anticipated benefit, or are even practicable. The changes should, when implemented, be tested against prototype again to see whether they are indeed what the User intended.

When a decision has been made to implement such changes, the team leader must pass details back to the relevant analysts to investigate the ramifications of such changes.

The analysts, having carried out their investigation, will pass details to the Stage 3 technical manager (assuming the appropriate Project Management structure) for approval of the changes, or otherwise.

One aim of Specification Prototyping is to identify new requirements, as well as to validate current ones. If this happens, the team leader can update the Requirements Catalogue directly, at the conclusion of the prototyping activity.

Possibly the prototyping activities in Step 350 will result in more functions being identified, for example. If this happens these should be given a Function Definition of their own.

Confirm the specification content

When all the prototyping has been completed, all the changes recorded and implemented, the Requirements Catalogue amended, the team leader must prepare the report to management on the exercise. This report will cover several related issues:

- Have all the nominated dialogues and reports been prototyped?
- Have all the prototyping objectives been fulfilled? If not, why not?
- How has the Requirements Specification changed as a result of the prototyping exercise? This may be through new requirements, a clarification of identified requirements or an acknowledgement that earlier requirements were not needed after all. Which SSADM products have been changed, and how, as a result of the exercise?
- What lessons have been learned about prototyping from the exercise? Was the exercise worthwhile? Should it be approached differently next time?

This report will be the major output from Step 350, along with any amended SSADM products.

SUMMARY

Specification Prototyping is not a method for building a system; rather it is a way of validating the agreed requirements by showing the user an animated model of them.

This has a number of advantages: it enables the user to get an idea of what the system will look like; it acts as a generator of ideas for new requirements; it identifies faulty analysis in time to be corrected before design is started; most importantly, perhaps, its reliance on the User as participant ensures User commitment to the new system.

Specification takes place in Step 350, following Function Definition and Dialogue Definition. Its output is a report to management, an amended Requirements Specification and, sometimes, amendments to other SSADM documentation.

16. Requirements Definition

16.1 Aims of chapter

In this chapter you will learn:
- How SSADM describes the requirements for the new system.
- How and when SSADM gathers information about functional requirements and non-functional requirements.

16.2 Where Requirements Definition is used in SSADM

The main tool of Requirements Definition, the Requirements Catalogue, is created and amended in the following steps:

Step 110—Establish analysis framework   The Requirements Catalogue for the Full Study is created in this step. If a Feasibility Study is carried out, then a catalogue for that study is carried forward.

The first requirements entered are those identified on the PID, or equivalent document.

Step 120—Investigate and define requirements   This is the most significant early step for the Requirements Catalogue. If there is a current system, the step is undertaken in parallel with Steps 130 and 140, in which the current environment is studied; if this is a greenfield site, this step is performed alone. Whichever circumstance applies, the products from Step 120 drive the Business System Options in Stage 2.

The entries from this Step combine fresh requirements from the new system with identified problems in the operations of the current one.

At this point, the emphasis is on functional requirements, but a record may be made of certain non-functional requirements, such as security requirements, major constraints, etc., to aid in the creation and selection of Business System Options.

Step 150—Derive logical view of current services   When the current view of processing is logicalized, physical constraints that contribute to identified problems may be resolved. If that happens, then record the facts in the Requirements Catalogue.

Step 210—Define Business System Options   The Business System Options are identified from the entries in the Requirements Catalogue. Suggested solutions to the problems and requirements that feature in the BSOs should be recorded.

Step 310—Define Required System Processing }
Step 320—Develop Required Data Model }

The selected BSO may not address all of the requirements previously identified. The Requirements Catalogue should be annotated to explain which requirements have not been met, and why. It is possible that a later stage, e.g. Technical System Options, they may be addressed after all.

Expand the Requirements Catalogue in detail during these two steps, entering details of new requirements that may surface. Cross refer to DFD and LDS elements that support particular requirements.

Non-functional requirements, such as retention requirements, access, security requirements, etc., are described in Step 320.

Step 330—Derive system functions   Update and enquiry requirements are annotated in the Function Catalogue in this step. Cross refer Requirements Catalogue entries to the Function Definitions. There will not necessarily be a one-to-one correspondence between requirements and functions.

Step 350—Develop Specification Prototyping   In this step we sit down with the Users and make use of the prototyping techniques to clarify, and improve, their understanding and ours of the requirements.

If further requirements are identified during the process, these are entered into the Requirements Catalogue.

Step 370—Confirm system objectives   The Requirements Catalogue entries are thoroughly reviewed, along with the Required System LDM and Function Definitions, for completeness and accuracy. In this exercise we are making sure that all requirements, especially the non-functional requirements, have been identified and described fully.

Step 410—Define Technical System Options   When we reach this step, most of the requirements to be met by the new system will have been identified and addressed. However, there will still be some that have not been met by the logical model, but are more properly in the domain of Technical System Options. Service levels and technical requirements across the business system may come into this category.

Step 510—Define User Dialogues   Any requirements to do with dialogues are resolved in this step. See Chapter 14 for a fuller explanation of the related requirements.

Use in SSADM

Requirements Definition is one of the driving activities of SSADM. The emphasis of the method is on the future system, whether it is to replace a current system, or whether it is a greenfield site.

There is no one rigid Requirements Definition technique to describe; rather it is a continuous process that involves constant User consultation, which is carried out from the beginning of the project up to the end of Logical System Specification.

16.3   Relationship with other techniques

As stated earlier in this chapter, Requirements Definition is not a discrete technique like, for example, Data Flow Modelling. Rather, it uses a variety of skills to identify the requirements and enter them into the Requirements Catalogue.

Any SSADM technique can use the Requirements Catalogue entries, or cause amendments to be made: it is the project repository of requirements information.

The particular SSADM techniques associated with it are:

Data Flow Modelling   see Chapter 9

Function Definition   see Chapter 11

Logical Data Modelling   see Chapter 10

Business System Options   see Chapter 17

Technical System Options   see Chapter 17

Specification Prototyping   see Chapter 15

As well as the core SSADM techniques, Requirements Definition also has strong links with the non-SSADM techniques outlined in Project Procedures (see Chapter 2).

The principal procedure techniques related to Requirements Definition are:

● *Capacity Planning*   This is required to ensure that there is sufficient capacity to meet the application's requirements. Also, it is necessary to ensure that meeting the new application's requirements will not seriously degrade current services.

Using Capacity Planning techniques we can paper test the required service levels against suggested technical environment descriptions.

● *Risk Analysis and management*   This technique is intended to identify and guard against likely security threats to the information system, whether from terrorist threat to the premises, from 'hackers' illegally accessing the data, from fire threat or from data corruption.

Requirements Analysis should identify those areas of vulnerability that are most likely, and interact with the Risk Analysis techniques to see best how to meet them. There is no fixed step in SSADM where security and control considerations must be addressed, but the questions must be posed in parallel with the SSADM project.

● *Testing*   Although physical testing of the system occurs after SSADM's involvement with the project has finished, the Requirements Catalogue makes testing objectives easier. Every requirement, whether functional or non-functional, must have a quality criterion which is measurable and quantifiable. These criteria must be entered in the Requirements Catalogue, and form the basis for subsequent test design.

- *Training and documentation*   The analyst must be aware of the need for the User skills to be developed if the system is to work, however technically excellent it may be. The two tools to develop User understanding are training and clear documentation. Requirements of Users and support staff must be noted and addressed.

16.4   Defining the requirements

This section describes the activities involved in Requirements Definition, including the Requirements Catalogue, and the distinctions between functional and the different kinds of non-functional requirements.

Identifying requirements

Another technique applied to requirements is that most basic skill of the systems analyst: fact-finding. This book is not an appropriate place to describe the various fact-finding techniques, but will point out the features that the analyst is aiming to elicit with reference to requirements.

When carrying out the fact-finding exercise, the analyst must highlight the following points:

- What is required from the new system?
- Which User will 'own' this requirement?
- Why is it required? Is it vital, useful or a nice idea? Some projects will fail if a particular requirement is not implemented, and be unaffected by others.
- What measures can be applied to the requirement? Do not accept a requirement that cannot be quantified by some appropriate measure. This measure may not be forthcoming at first identification; if that is the case, record the requirements in the Requirements Catalogue and return to the question of measurement later on.

Requirements Catalogue

This document is the repository of all requirements information. The first entries are made during Feasibility, and carried forward to the start of the Full Study.

All entries are regarded as provisional, and should be returned to, and revised, as often as need arises up to the end of Logical System Specification.

The emphasis should always be on the future system, rather than on the current one. If there is no current system, the first requirements entered here will form the basis of Logical System DFDs, LDMs and BSOs, as envisaged by the users of the new system.

Figure 16.1 illustrates an entry for the Requirements Catalogue for SS plc.

Requirements Catalogue Entry

Source Booking Officer	Owner Course Mgr.	Requirement ID 9	Priority H

Functional requirements

Provide Booking Clerks with on-line access to names of delegates on Waiting List for a given course, to speed up replacement of place after delegate cancellation.

Non-functional requirement(s)

Description	Target value	Acceptable range	Comments
Response Time	3 seconds	3-6 seconds	
Service Hours	9.30 - 5.00 Mon - Fri		
Availability	90%	85-90%	

Benefits

Will speed up process of finding names and creating an offer. Will free Booking Clerks for other pressing work.

Comments/suggested solutions

To fit terminals on to two clerks' tables, with on-line access to Waiting List names, in first-come, first-served order, or in Branch order.

Related documents

Required System DFD, Process Box 5

Related requirements

7. To offer places to Delegates from same Branch as Delegate who cancelled.
11. To identify cancellation penalties

Resolution

Figure 16.1

Functional requirements

These are the requirements that, from the user's point of view, perform the activities that run the business. These will include all updates to master files, enquiries against the data on file, producing reports and communicating with other systems relevant to the business activities.

Non-functional requirements

These requirements define the performance levels of the business functions. These will include such features as response times (for online transactions), turn-round time for batch inputs and levels of accuracy. Other non-functional requirements concern such issues as security, recovery and back-up in case of breakdown.

Some of these may be common across the whole business system, such as back-up facilities, while others may be specific to a given application, such as response time. Whether the requirement is local or global should be noted in the Requirements Catalogue entry.

The non-functional requirements are likely to be reviewed and revised several times during the study as the analyst learns more and more about the system. Step 350, Develop Specification Prototyping is a likely place for amendments, as the User sees precisely what is offered, and understands what is possible.

Quantifying requirements

To avoid ambiguity, and to give a basis for testing, requirements should be given some form of quality measure. The act of quantifying the requirement helps to focus the analyst's attention on the specific requirement. Vagueness suggests that the requirement may not be fully understood.

This is obviously easier in the case of non-functional requirements. Some functional requirements may not lend themselves to quantification easily. For instance, if a requirement is to improve customer satisfaction, how do we measure something as intangible as that? We must settle on some criterion that tells us that has been achieved. A possible measure is that we receive a certain proportion of verbal or written compliments, or that our business with present customers increases by a set percentage.

Specifying requirements

The entries in the Requirements Catalogue express the User's perception of what the new system is to achieve. They are not precise enough, though, to act as a specification for the new system.

Each entry needs to be supported by the more rigorous expression of the SSADM techniques of Entity/Event Analysis, Logical Data Modelling, and Function Definition. These will all be based on the entries in the Requirements Catalogue, and can be traced back to the appropriate entry, but are more precise than the textual descriptions in the entries.

Entries which are not met by these techniques should be carried forward into specification. They may be addressed by other techniques, or by the particular technical implementation.

SUMMARY

Requirements Definition is a continuous process, not a simple step activity. It is the driving force in a SSADM project, always forcing the emphasis onto the new system, rather than the current environment.

The principal tool in Requirements Definition is the Requirements Catalogue. This is started at Project Initiation, and is still updated even during Physical Design.

From the first steps, we are identifying the functional requirements and, to an extent, some non-functional requirements. The latter are examined in more detail as the study proceeds.

All requirements should be quantified to provide a basis for planning and evaluating tests later in the cycle.

Participants in Requirements Definition will be the analyst, designer, User and also the service providers, who will define the non-SSADM requirement techniques such as Capacity Planning and Risk Assessment.

17. User options

17.1 Aims of chapter

In this chapter you will learn about the three places in SSADM where the User is presented with a selection of options, and must choose which way the project will progress.

- Step 030   Identify Feasibility options
- Step 210   Define Business System Options
- Step 220   Select Business System Option
- Step 410   Define Technical System Options
- Step 420   Select Technical System Option

17.2 Use in SSADM

The options are of two basic kinds: Business System Options, which provide us with the scope and objectives of the project, and Technical System Options which define the technical environment, in terms of hardware, software and development approach.

In each case, the analyst prepares a set of options with full supporting documentation and Cost/Benefit Analysis. The User, after the presentation of each option, considers the relative merits of each, and makes a selection. This choice could be to abort the project at that point, or to proceed along a certain path.

17.3 Feasibility Module

The Feasibility Study is usually instigated as a part of a Strategic Study carried out some time earlier. This Strategic Study will have identified areas for computerization and redesign, and accorded priorities to the various project areas.

The objectives of Feasibility are to investigate the project area and identify (a) whether or not the project is technically feasible, and (b) whether a sound business case can be made for pursuing it.

The end of Feasibility is a set of options presented to the Project Board, each making a different recommendation about the path to be followed.

It is possible that if circumstances have changed since the Strategic Study, that no good business case can be made for pursuing it. If so, the recommendation will be to abort, or at least postpone, the project there.

Each option meets the requirements defined in the Project Initiation Document and identified in the earlier parts of the study, at least to a specified minimum level. The analyst will also produce for each option outline project plans as to how to proceed.

Business System Options and Technical System Options, are described in detail later, so I shall not dwell on them here, other than to say that, at Feasibility, each option will contain elements of Business System Options and Technical System Options.

Feasibility Options are high level only, and far less detailed than the full Business and Technical System Options. The outcome of Feasibility is a recommendation to proceed, along the lines suggested, but a Full Study will follow, and that is where the detailed analysis of the costs, benefits and impacts of the options takes place.

17.4   Business System Options (BSO) — Requirements Analysis Module

Stage 2, in Requirements Analysis, is concerned with the selection of the Business System Options. The two steps cover all BSO activities, and form the major input to the next Module, Requirements Specification. A BSO describes a suggested new system in terms of its functionality and its boundary. Inputs, outputs, processes and data are described, just as in the Current Environment Description. Its aim is to help the Users choose, from all the listed requirements, just what they want their new system to do.

Its input is the Requirements Catalogue and Current Environment Description. Development of the BSOs is done in consultation with the Users. When the Users are presented with the options there should be no surprises.

The format of the BSOs is a text description of the boundary and functions to be performed, although these will probably be illustrated with high-level DFDs and an LDM. At option development these techniques are not to be used in detail, but rather to give a broad-brush view of the system. They are considerably enhanced when the selected option is built up into the specification of requirements.

Technique

First, draw up a list of about six BSOs, covering a range of the requirements identified in Stage 1. The range should cover:

- one option that covers the stated minimum requirements, and no other;
- one option that covers every new requirement;
- up to four options that each cover the stated minimum requirements and a different set of the other requirements.

The six options will then cover six different boundaries and six different functionalities, yet all will cover the essential requirements identified as necessary for the new system.

These first options should be skeleton only, stating nothing more than what the option is going to cover. They should then be expanded to describe the impact upon the business, and the non-functional requirements that have been met. The impact upon the business should be expressed in terms of the priority of the BSO in relation to the IS strategy.

Second, as the BSO is expanded further, bring in details of expected volumes: volumes of data; volatility of data; frequencies of key tasks, especially at peak times.

Also describe the BSO in terms of project development, looking at the following aspects:

- cost/benefit of proposed option;
- impact analysis of implementing the BSO;
- timescales for development and construction.

Finally, when the six skeletons are produced, the analysts and Users together must begin eliminating obvious non-starters. The criteria may be cost, or organizational impact or inadequate functionality. Whatever reason, the list should be reduced to two or three options for the presentation.

Having whittled down the list, we now expand the remaining options with greater detail. The prose description of the options will be longer and more explicit, defining more closely the functions performed, inputs/outputs, data and processes. The selection of processes for batch and online processing are also defined, with a (rough) idea of service levels that can be provided.

The selection considerations are defined in greater detail. These take into account:

- costs/benefits,
- constraints,
- impacts on existing systems,
- plans/timescales for subsequent SSADM activities, and implementation of the system, and
- organizational impacts and implications.

The differences between the individual options are unlikely to be radical. They may depend on a trade-off between cost, security, functionality and service levels, with seemingly marginal differences between any two. This is obviously acceptable, as long as the differences are not so marginal as to be not real.

Selection of BSO

In Step 210 we have, with the Users' help, defined the options from which they will choose. In Step 220 we present the options to the body that will decide on the future course of the project.

The presentations may be made to the Project Board directly, or to a User body that has been empowered by the board to make the decision.

Standard presentation techniques, not specific to SSADM, are employed by the analyst for the exercise. The four basic tasks are:

1. Prepare the presentation.
2. Deliver the presentation.
3. Help the User decide, and afford clarification as needed.
4. Record the selections made.

At the presentation, the analyst will identify the points of difference between the shortlisted options. The review body will probably want to question the team further about distinctions in functionality, service and response levels, costs and development times. With the answers to these points the Users may then make the decision.

The decision may be to accept one of the options as it stands, or to make a hybrid selection, combining features of two or more options. If this happens, the analysis team must carry out a fresh exercise of costing, justifying and scoping the new option. This will be done in greater detail than the others, as it will form the basis for the specification for the new system design.

Another decision may be to stop the project at that point, or to reject all of the options and ask for more to be prepared. If that happens, it suggests that the analysts did not carry out the preparation thoroughly enough; by the time that presentations are made, the Users should be familiar with the likely courses to follow and should have made it clear if none was acceptable. However, there may be other sound business reasons for rejecting at that point. If it happens, the analysts must backtrack and examine the Requirements Catalogue again.

Whatever decision is taken, it must be recorded in the project documentation. Details to include are:

- option chosen,
- reasons for adoption,
- options rejected, and
- reasons for rejection of each.

These details, plus a detailed description of the Selected BSO will be passed, via the information highway, to Stage 3.

Application of BSO

As with Technical System Options, what I have described is a set of guidelines to help the analysis team to reach a solution to the requirements. They are not intended to be fixed rules, or an algorithmic technique. Many SSADM techniques are 'soft' rather than 'hard', and BSOs, being creative and forward-looking, are necessarily very soft. If this means that your organizational standards dictate a different approach, then follow that.

What SSADM does stress is the role of the User in choosing, with the analyst's careful advice, the business and functional path that the project is to follow.

17.5   Technical System Options (TSO) — Logical System Specification Module

Stage 4 is concerned with the production and selection of Technical System Options. TSOs are the decision point after BSOs, where the Users are able to decide the future course of the development, even to the point of abandoning the project.

The TSO defines the implementation environment and strategy for the Selected BSO. The issues addressed are:

- Specification and description of the hardware, software and data environment. This will be in generic terms, and will not lead to a tie into any particular vendor: tenders will not be invited at this stage.

- Confirmation of the functions in the application area, and mode of processing for each.
- Description of the organizational impact and effect on work methods.
- Description of the impact on the rest of the development organization, and the remainder of the project.

Many of these issues may have been addressed already by the Feasibility Study, and/or the IS strategy for the given functional area. If this is so, then the range of TSOs is constrained by these earlier decisions, which are to be found recorded in the Requirements Catalogue. If there has been no Feasibility Study, and the IS strategy has not tied the system into a particular technical environment, these issues are all considered.

Inputs to TSO are a mix of SSADM and non-SSADM documents.

SSADM inputs are as follows:

- Requirements Specification
- Selected BSO (with reasons for selection)

Extra-SSADM inputs are as follows:

- Project Initiation Document (PID)
- Description of current IS environment
- IS strategy
- Standards
- Security standards
- Installation Style Guide

While the SSADM inputs are the drivers of TSO, the other management inputs are a source of influence and guidance.

The options themselves take the form of:

- Outline Technical Environment Description (TED)
- System Description
- Outline Development Plan
- Cost/Benefit Analysis
- Impact Analysis

Each of these is completed in outline only: with several options to complete, only one of which will go forward, it is too much effort to give full and detailed descriptions for each.

Technique

The procedure for producing and selecting TSOs is very similar to that for BSOs. The difference lies in the objectives of the options, and the constraints that apply. This section describes the techniques briefly, and then looks at the particular constraints that affect technical, as opposed to business, options.

First, draw up an initial list of six or so options. These may come from methodical discussion of different approaches, or may be generated by a brainstorm. However they are raised, once the six are created in skeleton form, they must be expanded. This expansion though, often requires investigation of suppliers, to obtain details of such things as costs, facilities, performance, support, etc. Note: this is not to pre-empt choice of vendors, but to obtain 'ballpark' figures and estimates to present to the Project Board for each option.

If the current system is a manual one, then one of the initial options must be a no-computer option. Similarly, if the current system is a computer-based system, one option to be considered should be a no-change option, i.e. end the project here.

Second, six TSOs are too many to specify in sufficient detail for presentation. We must therefore cut them down to a more manageable number.

Three is an acceptable number, but you may find that you want or need to include a fourth viable option. Often this can be done automatically, as some options are obviously more appropriate than others. Always consult with the User, on an informal basis, as to which should be taken forward. It may be that they ask for features from two separate outlines to be combined into one new option. Any feature which is not acceptable or appealing to the User can be identified and discarded at this point.

Third, once the number of TSOs has been agreed, the skeletons must be fleshed out with detail. The proposed configuration of each will be the central document.

Aspects such as required data volumes are needed to assess the viability of a suggested hardware/software configuration, so a rough sizing exercise for each option must be carried out. A non-SSADM technique, Capacity Planning, will help in this task. Capacity Planning is one of the Project Procedures described in Chapter 2.

As the options are developed, changes may be needed to earlier documents, or changes made to the options in the light of earlier work in the project. Some form of change control mechanism should be set in place, so that changes can be implemented and recorded without inconsistencies and inaccuracies arising.

Changes to earlier documents might take the form of amending ELHs to assess or alter supervisory events and constraints, or changing/creating entities on the LDS to take account of the proposed configuration's capabilities/constraints. Other products may be susceptible to change during the TSO development. The information highway must be able to pass the changes to the relevant control authorities to preserve the system's integrity. It may even be that, as the Selected TSO is being refined and expanded later, a complete second pass through Stage 3 is required to ensure that the design process has full information to work on.

Fourth, the options to be presented are prepared according to installation standards, but must contain certain information to allow the Project Board/User body to make a sound decision. The features that must be included are as follows:

OUTLINE TECHNICAL ENVIRONMENT DESCRIPTION (TED)
This describes the hardware, software, development environment, system size, and fallback/recovery proposals. The TED explains to the User how the system works, and how it will be developed. It should contain enough information to allow reasonable costings to be made. At this point, however, the TED is in outline only. After selection, the chosen TED will go forward as the major document.

SYSTEM DESCRIPTION

This may have been covered already in the BSO. Its province is the solution of items in the Requirements Catalogue. To reduce effort in the preparation of the TSO, you may wish to use SSADM products such as the Function Definitions, annotated Requirements Catalogue, and Required System DFM and LDM.

IMPACT ANALYSIS

This is concerned with the impact of the proposed system on the User's environment. The issues involved are general Systems Analysis and Project Management issues, not particular to SSADM. The concern is not with the technical feasibility of the technical system, but with the organizational implications. The issues involved cover the following topics:

- organization and staffing,
- changes in operating procedures,
- savings from replaced equipment, or from maintenance no longer required,
- implementation considerations (from staffing point of view),
- training requirements,
- User-manual requirements,
- testing requirements,
- take-on requirements and
- advantages/disadvantages against the other TSOs.

OUTLINE DEVELOPMENT PLAN

This plan provides management with proposed strategies for developing the particular options, with ideas of timescales and resources required. From this information, management can prepare plans and compare the development overheads of the respective options. The plan should contain information under the following headings:

- system design,
- program design/coding/testing,
- procurement (if relevant) and
- implementation.

The precise contents of these sections depend upon the circumstances and standards prevailing at the installation. For each, though, some statement should be made that will enable management to carry forward the planning and costing of development into the next stage.

For the estimating exercises—both time and cost—the estimating subject guide referred to in Chapter 2 will be necessary, unless local estimating methods are prescribed.

COST/BENEFIT ANALYSIS

This provides the Project Board with quantifiable selection criteria. It deals with the financial specification of each option, and so is the part that often decides the board's choice.

It covers a number of areas, under both costs and benefits, the principal ones being:

- *Development costs, or non-recurring costs*   These can be calculated from the TED, after consultation with a number of vendors. The outline development plan provides costs of resources needed for the development.
- *Operating costs, or recurring costs*   The TED and Impact Analysis provide data for calculating these.
- *Tangible benefits*   These are measurable financial benefits of implementing the new system. Examples are money saved in better stock control, or higher interest from improved cash flow, or increased profit margins.
- *Intangible benefits*   These are benefits that are harder to quantify, but result from implementing the new system. Such things as improved customer service, higher staff morale and better work environment are included under this heading. While a precise figure cannot be put on such benefits, some yardstick for measuring should be attempted, such as adding one percent business for satisfied customers, lower staff turnover or saving X per cent recruitment costs per annum for higher morale.

Selection of TSO

The Procedure here is the same as that for BSOs. There are four tasks involved:

1. Prepare the presentation.
2. Make the presentation.
3. Assist the Users/Project Board.
4. Record the selection decision.

As with the BSOs, the final selection may prove to be a hybrid of features from two or more options. It is unlikely that a decision will be made immediately the presentation has finished: there will be too many details to consider. However, a date in the near future should be agreed for a decision, so that the development plans, and therefore costs, are not impacted by any delay.

The decisions, should be recorded, and the TSO and TED updated accordingly. The TSO documentation is filed away now, and the TED of the selected option expanded to be carried forward to Physical Design.

The selected option is examined again from the point of view of Capacity Planning to make sure that it meets the service-level requirements. If they cannot be met, another decision must be made:

- Propose a new architecture with greater capacity.
- Reduce the service-level targets.
- Propose alterations to the Requirements Catalogue.

Whichever body accepted the option must meet to decide upon the choice to make here, and to decide what extra work needs to be undertaken.

SUMMARY

There are three places where the Users take a crucial stop–go decision in a SSADM project:

1. Feasibility (Stage 0),

2.   Business System Options (Stage 2) and
3.   Technical System Options (Stage 4).

The Feasibility Options include elements of BSO and TSO, and may set the decisions for the subsequent project development.

BSOs give the User a set of options that define the scope and functionality of the new system.

TSOs give the User a set of options that show how the BSOs are implemented. This is the option point that provides the planning and costing data for the Project Board. The development of the TSOs may, under some circumstances, cause a second pass at Stage 3, so requiring a Configuration Management procedure to be set up.

The output from TSO forms the input to Physical Design at Stage 6.

18. Logical Database Process Design

18.1 Aims of chapter

In this chapter you will learn:
- How to derive and draw Enquiry Processing Models.
- How to derive and draw Update Processing Models.
- How to recognize structure clashes.
- How to classify success units.

18.2 Where Logical Database Process Design is used in SSADM

Logical Database Process Design (LDPD) is performed in the following steps:

Step 520—Define Update Processing   In this step we complete the Entity Descriptions with state indicators. Entity Life Histories and Effect Correspondence Diagrams are converted, using Jackson diagramming techniques, into structure diagrams that reflect the processing logic of given events.

Step 530—Define Enquiry Processing   The Enquiry Access Paths from LDM are taken as the input structure to the enquiry process, and the I/O Structure from the Function Definition is transformed into an output structure. The two structures are merged to form the enquiry process.

Use in SSADM

LDPD is used to define the processing requirements on the Logical Data Model. It translates the information gathered during Requirements Specification into a processing logic which is implementation independent, and is so well defined that it provides ease of maintenance after implementation.

The Logical Process Specifications are fed into Stage 6, Physical Design. If a 4GL is being used to generate the code, the specification from LDPD should be sufficient input; if a 3GL environment is to be used, Physical Process Specification will translate the Logical Specification into a form that can be coded to access the physical database.

The Process Models consist of:

250

- a diagram, either Enquiry or Update Processing Structure, and
- an operations list as supporting documentation.
  Inputs to LDPD are as follows:
- Effect Correspondence Diagrams
- Enquiry Access Paths
- Entity Life Histories
- Function Definitions
- I/O Structures
- Required System Logical Data Model
  Outputs from LDPD are as follows:
- Enquiry Process Models:
  —Enquiry Process Structure Diagram
  —Enquiry Process Structure operations list
- Update Process Models:
  —Update Process Structure Diagram
  —Update Process Structure operations list

Figure 18.1 illustrates the relationship of LDPD with other SSADM techniques.

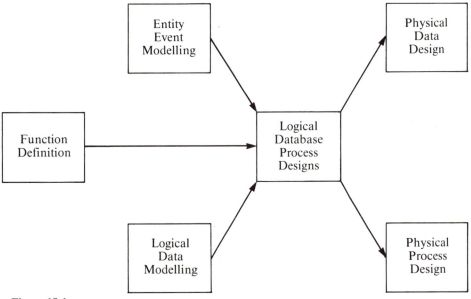

Figure 18.1

18.3 Procedures

Figure 18.2 depicts a Universal Function Model, i.e. a model that shows the different components of a function, from input to output processes.

LDPD is concerned with those parts that are associated with the Update or

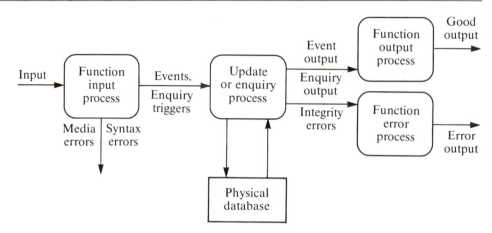

Figure 18.2

Enquiry Processing components. The processing requirements of the business functions are translated into the standard sequence/selection/iteration structure, together with the operations that are carried out to perform the function, or, usually, subfunction.

The parts of the Universal Function Model that handle the media/syntax errors on input, or integrity errors on database processing, and outputting error reports are dealt with in Stage 6, Physical Design.

Enquiry and update processes are approached in similar ways; one through the Enquiry Access Paths, the other through Effect Correspondence Diagrams. They seem at first sight to be independent, but in fact may prove to be closely connected, in two ways:

1. An update process may need to be extended to produce an output report. If so, the enquiry processing procedures will need to be followed.
2. An update process is often introduced by an enquiry process. You must investigate first to see if such an occasion leads to two success units, or just the one. The answer will tell you whether the two processes are to be treated as one unit or two.

Enquiry processes

The basic description of the procedures to follow is simple: unfortunately, simplicity is not often to be found in the information world, and so the application of these procedures leads to a number of possible complications. I shall begin by describing the simple procedures.

Enquiries come in two basic forms: structured, enquiry functions, and ad hoc enquiries, which are more loosely specified. Ad hoc enquiries are, by their nature, hard to model, and often hard to anticipate; I shall therefore only look at the formal enquiry functions here.

Specify the enquiry name   Each enquiry must be uniquely identified, the name being used also in Function Definition and Physical Process Specification.

Specify the enquiry trigger   The trigger consists of the data items input, as identified in Logical Data Modelling Enquiry Access Paths. If the enquiry requires a simple navigation around the data model for specific entity occurrences, the trigger will be the key of the entity which is the entry point.

If, on the other hand, the enquiry is to find all occurrences of a class or category, such as all tutors not assigned on a certain date, or all branch managers due to be invoiced, then the trigger will be the name of the data items and the search criteria, which will be the specified values for those items.

Specify the Enquiry Access Path   In Step 360 we created Enquiry Access Paths from the LDS, one for each specified enquiry. Figure 18.3 is the Enquiry Access Path for the requirement, Generate Course Timetable.

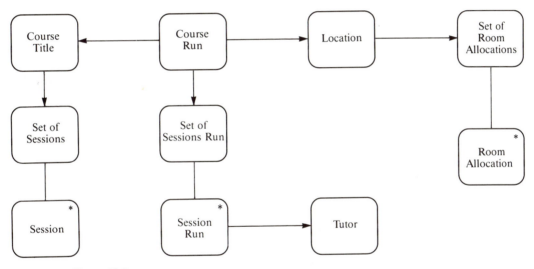

Figure 18.3

Specify the enquiry output   In Step 330 we completed I/O Structures for identified requirements. Figure 18.4 gives the I/O Structure for the same requirement, Generate Course Timetable.

Group accesses on the Enquiry Access Path   Group the accesses on the Enquiry Access Path that are in a one-to-one correspondence. Figure 18.5 shows this for our enquiry.

Convert to structure notation   Convert the Enquiry Access Path diagram to structure notation. It may be that new nodes must be inserted to conform with the rules of the structure model. Figure 18.6 shows the structure derived from the grouped Enquiry Access Path.

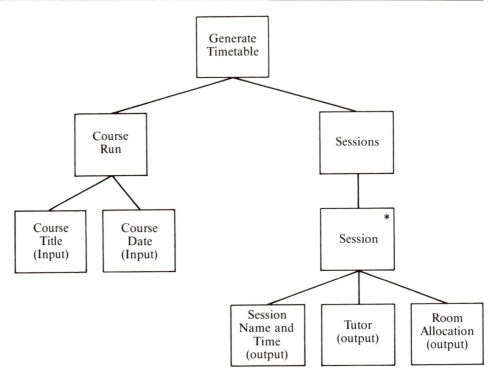

Figure 18.4

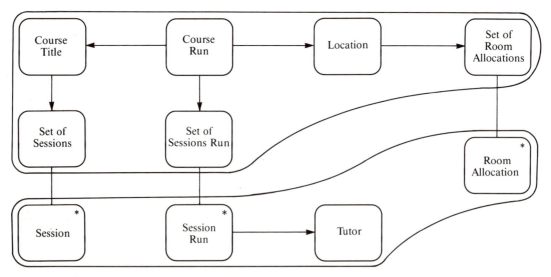

Figure 18.5

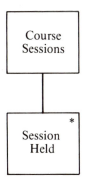

Figure 18.6

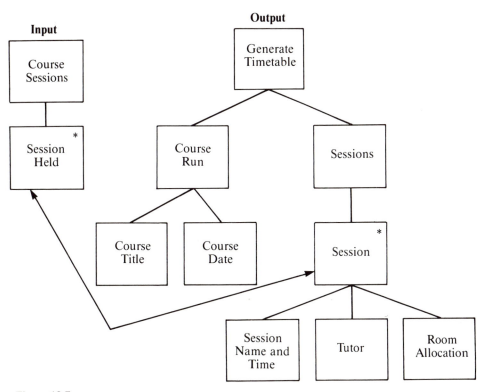

Figure 18.7

Identify correspondences between input and output data structures   The Enquiry Access Path is the input structure to this procedure, and the I/O Structure is the output structure. They have points of one-to-one correspondence on them. Mark these points with correspondence arrows. In Fig. 18.7 we see the two structures, and can see where the correspondences naturally occur.

Merge the input and output data structures   Enquiry Process Modelling consists of taking the input and output structures and merging them, according to the correspondences between them, as shown in Fig. 18.8.

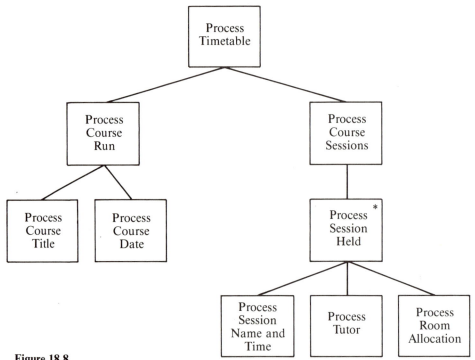

Figure 18.8

List the operations and allocate them to the structure   The operations that can be attached to the structure are, obviously, 'read' operations. Unlike the operations on the ELHs, they can be put under structure boxes as well as elementary boxes.

The permitted operations are:
- Read ⟨entity-type⟩ by key: read the database entity using the input key value.
- Define set of ⟨entity-type⟩ matching input data: define a set of ⟨entity-type⟩ entities, the members of which match the criteria in the input data.
- Read next ⟨entity-type⟩ in set: read the next entity of specified type from the currently defined set. There should always be a 'define set' operation before this one is used.

- Read next ⟨detail⟩ of ⟨master⟩ [via ⟨relationship⟩]: read the next detail (specified entity type) of current occurrence of master (specified entity type). The optional 'via ⟨relationship⟩' applies if there is more than one relationship between the two entity types.
- Read ⟨master⟩ of ⟨detail⟩ [via ⟨relationship⟩): read the master (specified entity type) of current occurrence of detail (specified entity type). The optional 'via ⟨relationship⟩' applies if there is more than one relationship between master and detail.

Figure 18.9 shows the operations list and the allocations for the Generate Timetable Enquiry structure.

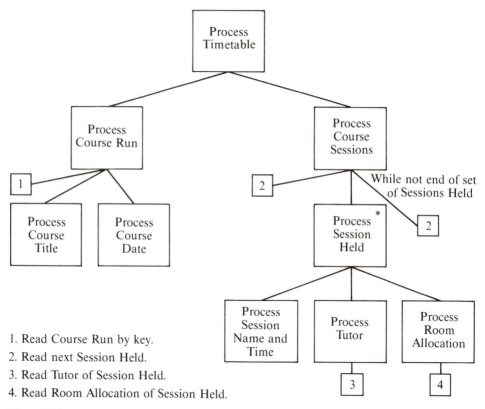

1. Read Course Run by key.
2. Read next Session Held.
3. Read Tutor of Session Held.
4. Read Room Allocation of Session Held.

Figure 18.9

Specify integrity error conditions   Integrity problems are usually associated more with update processes, but enquiries are obviously only valid if the database is correct so the errors must also be noted here.

The commonest kind of integrity failure that can be tested is the absence of a record that should be there. The record may not be in the valid state for the particular enquiry, or indeed, it may not be on the database at all. If that is the case, then there should be a ⟨fail-if . . . ⟩ operation built in to the operations that access the database. These should be shown as operations on the list, and inserted immediately after every database access.

Specify error outputs   With the User's help, design the error report, whether printed or displayed. It is up to the User whether the reports form part of the valid output report, or form a separate report on their own account.

Walk through the structure   Again, with the User, conduct a structured walk through of the Process Model, to ensure that it meets the requirements, and can actually retrieve all the information necessary to answer the enquiry.

Update processes

Our input to this is the Entity/Event Modelling. Every identified event will have an update process associated with it. Whereas enquiry processes are derived from Enquiry Access Paths, update processes are derived from Effect Correspondence Diagrams.

The source of the I/O Structures is more complex than the enquiry processes, in that each event may be defined in more than one function. Each Function Definition must be checked to ensure that the data items for the event are contained in the input structure.

Specify event data   The event data means attributes—normally the key—which act as the entry point to the Logical Data Model, and any additional updating information. This additional information may be new values for current attributes, or new key and attribute information for insertion of a new record.

Specify the Effect Correspondence Diagram   This is carried out in Entity/Event Modelling in Stage 3. It encompasses the effects for each event. By 'effect', I mean identifying all entities that are affected.

Specify event output   This means all output data items from an event, excluding any error messages. The output may be trivial, such as an acknowledgement that processing has been successfully completed, or it may be in the form of some structured, necessary output, which needs to be modelled on the same lines as enquiry output.

Extend the ECD with enquiry-only entities   If, for an update to be performed, reference has to be made to other entities, include those reference entities on the ECD.

Group effects in one-to-one correspondence   This is described in Chapter 13. Draw a box around each set of entities in one-to-one correspondence, and award each group a meaningful name as a process. Note: there may be one or two entities that are not included in any such grouping. That is acceptable. They too should be given a

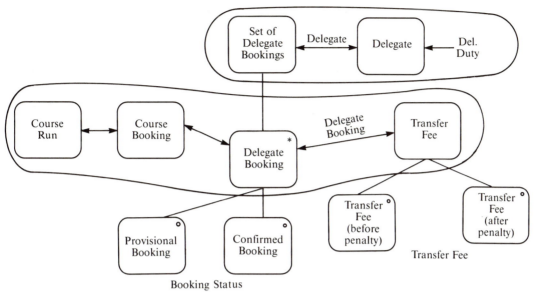

Figure 18.10

meaningful process name. Figure 18.10 shows the Effect Correspondence Diagram for the event Remove Delegate.

List operations For each entity affected, list the operations that are applicable. The classes of operations should, for each entity, be in the following order:

1. A read operation.
2. Operations to raise error messages on invalid state indicators. Invalid SIs are identified from the 'valid previous' SIs on each entity. Note: the SIs are added to the ELHs for each entity as the first task in the current step, Step 520.
3. Operations from the relevant effect on the ELH are expanded and added. 'Gain' and 'lose' operations are not carried forward. The ELH operations are expanded as shown below. The additions, which specify the particular occurrence affected, are shown in the curly brackets. Phrases in square brackets are optional.
   (a) Store keys {of ⟨entity⟩}.
   (b) Store ⟨attributes⟩ {of ⟨entity⟩} [using expression].
   (c) Store remaining attributes {of ⟨entity⟩}.
   (d) Replace ⟨attribute⟩ {of ⟨entity⟩} [using expression].
   (e) Tie {⟨entity⟩} to ⟨master entity⟩ [using ⟨relationship⟩].
   (f) Cut {⟨entity⟩} from ⟨master entity⟩ [using ⟨relationship⟩].
4. An operation to reset the entity's state indicator, using the valid 'set to' value.
5. An operation to write the entity.

There are some integrity checks upon operations:

- Every 'read' operation should be followed by an operation to check SI value, unless all values are valid for that effect.

- Every 'store key' operation should follow a 'create' operation.
- If any attribute, or the SI value of an entity, is changed there must follow a 'write' operation.

The operations list for Remove Delegate is as follows:

1. Read ⟨Delegate⟩ by key.
2. Fail if SI not valid.
3. Cut ⟨Delegate Booking⟩ from master ⟨Booking⟩.
4. Read master ⟨Booking⟩ of ⟨Delegate Booking⟩.
5. Fail if SI not valid.
6. Read master ⟨Course Run⟩ of ⟨Booking⟩.
7. Fail if SI < 5 > 6.
8. Replace ⟨No-prov⟩ of ⟨Course Run⟩ with ⟨No-prov − 1⟩.
9. Replace ⟨No-conf⟩ of ⟨Course Run⟩ with ⟨No-conf − 1⟩.
10. Cut ⟨Delegate Booking⟩ from master ⟨Delegate⟩.
11. Read detail ⟨Transfer Fee⟩ from ⟨Delegate Booking⟩.
12. Write ⟨Course Run⟩.
13. Cut ⟨Transfer Fee⟩ from master ⟨Delegate Booking⟩.
14. Set SI to null.
15. Write ⟨Transfer Fee⟩.
16. Write ⟨Booking⟩.
17. Set SI to (next valid value).
18. Read detail ⟨Delegate Booking⟩ of ⟨Delegate⟩.
19. Delete ⟨Delegate⟩.
20. Delete ⟨Delegate Booking⟩.
21. Delete ⟨Transfer Fee ⟩.

Convert to structure notation The ECD, with the effects grouped together, is converted into structure diagram notation. Figure 18.11 shows the ECD for Remove Delegate in structure notation.

Allocate the operations to the structure They should be allocated in the order indicated above in 'List operations'.

Allocate conditions to the structure Each selection and each iteration on the structure should have a condition governing the path to be followed. The condition may test the value of a state indicator or another value; it may test for the presence or absence of a record. Figure 18.12 shows the operations and conditions added to the Remove Delegate structure.

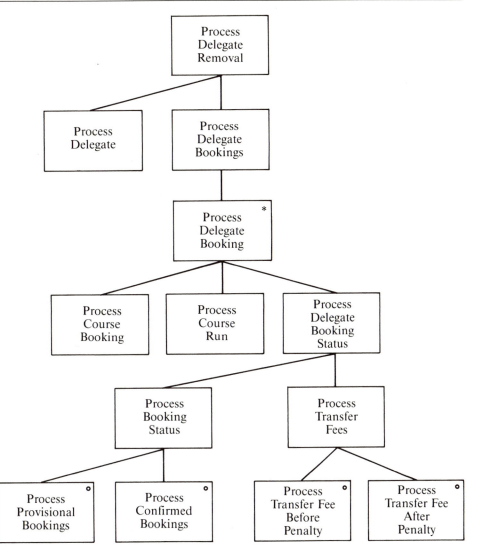

Figure 18.11

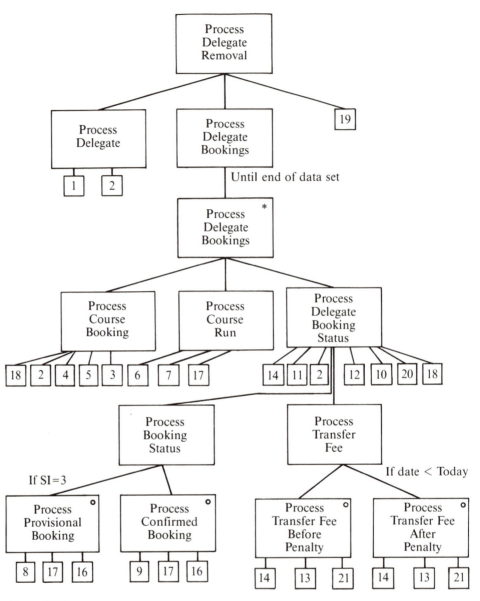

Figure 18.12

Specify integrity error conditions   Each update process must check that the relevant portion of the database is in a valid state before and after any updates are made. One kind of integrity check is the state indicator check made after any read operation. Other checks may involve examining a value of one attribute, or a combination of attributes, or relative values of two or more attributes. Such checks will usually involve applying business rules to the entity, so that some actions may not be taken if certain conditions are not fulfilled.

Each integrity error condition must be recorded on the operations list.

Specify any error outputs   The error outputs may be in the form of coded reports (i.e. giving a code number or letter to be allocated a meaning from a table), or an integrity error report. They may take the form of a separate report, a message to the screen, or a part of the standard output from the process. Whichever, it is a good idea to begin defining the error output at this point to prepare the way for Physical Design. It may even be useful to begin the Physical Design of the error output now.

Walk through the structure   This is a quality check on the product, to be held with the User, to ensure that it successfully models the Update Processing requirements. The quality checks on this mostly centre on the logical sequencing of the processing, and on the correct specification of integrity errors.

18.4   Processing features

The Enquiry and Update Processing steps described in Sec. 18.3 are portrayed as simple algorithmic procedures which inevitably result in an accurate model of the processes.

The world does not always work as neatly as that: there are two or three possible areas of complication in both forms of processing that I shall highlight here. I shall not describe in detail the ways of resolving all the complications: that is properly a subject for a publication on structured programming, although one of the CCTA subject guides to support SSADM (see Chapter 2) does deal with the commonest problems.

Structure clashes

When merging input and output data structures, the implication is that they will be very similar, and so correspondences are easily identified, and merging is automatic. In fact, frequently the two structures are dissimilar, and need to be resolved. This is termed a *structure clash*, and is classified under three possible headings: ordering clash, boundary clash, or interleaving clash.

ORDERING CLASH

This means simply that the data items in the input structure are ordered in a different sequence from the output structure. This is usually because of the sequence in which the data must be accessed on the LDM. There are two ways of resolving this:

1. Modify the LDM. This will be carried out in Stage 3, if it is the chosen solution. The model will be changed to allow more flexible access to the required data.
2. Define extra processes to sort and/or format the data. This is likely to be defined in Stage 5. The Universal Function Model (see Fig. 18.2) shows this as an output process.

BOUNDARY CLASH

Here, the data elements are grouped differently in the input structure from the output structure. This is because on the LDM the detail entities are grouped beneath the appropriate master, while on the output structure, the grouping is done on ergonomic principles, relating to the output display. There are, again, two solutions to this problem:

1. If you have an automated tool, such as an application generator or a report generator, let that create the output. It will resolve the clash according to the output specification.
2. Define two communicating processes, one for input and one for output. The output should be drawn to be compatible with the LDM structure. A separate formatting or sorting process can be created, as an output process, to handle the output.

INTERLEAVING CLASH

This happens in one data structure, the input. The data elements in the input structure represent more than one entity, but all are interleaved together, rather than all for the first entity, then all for the second and so on. To resolve this, Physical Process Design will create work space and extra processing to group the data elements in the optimum way for processing. See Chapter 20 for further explanation of this.

Common processing

During Logical Database Process Design, it will be found that several processes share common processing, perhaps in input and output handling, sort/formatting, or database access. When such common processing is identified a common Module should be created, and invoked. The Function Component Implementation Map in Stage 6 specifies how such common processes will be implemented physically.

Success units

When designing a logical database process the designer and User together must define the *success units* for each process, i.e. how much of the process must be performed for it to be deemed a success.

During the execution of a success unit, those portions of the database being accessed are effectively locked from other processes, so that the state of the database at the end is consistent with its state at the beginning of the process.

If anything should happen during the process to cause it to be abandoned— discovery of an integrity error, say, or the User who initiated an online task having to leave it for something else—the database will be 'rolled-back' to its state at the start of the process, cancelling any updates that were carried out up to that point. If the success unit is completed successfully, then the database will be 'committed' or 'rolled-forward' to a new state. In this case, all updates will be effected on the new database.

The designer must agree the success unit strategy with the User. There are a number of possible strategies open to them:

An update process is a logical success unit
In this strategy, which is more or less inevitable, each update process is a success unit. This is recognizing that an event has triggered the update, and all changes resulting from this event must stand or fall together, rather than being implemented piecemeal.

An enquiry process is a logical success unit
This strategy states that an enquiry is triggered by a specific input, and while it is being executed the whole database should be locked to preserve integrity. Thus the trigger is equivalent to an event, and all accesses must stand or fall together, in a similar way.

Offline functions will almost certainly be implemented this way; online functions may need to be broken down into smaller processes, or use extra Modules to cope with interleaving clashes or boundary clashes. The enquiry should not, however, be broken down into smaller success units.

An enquiry process may be divided into smaller units, where each is a logical success unit
This is a possible strategy for an online enquiry which accesses large amounts of data. The justification for this is to give the User a chance of abandoning the enquiry—for whatever reason—before the end, but without losing all the information gained up to that point.

It may be, for example, that an enquiry is to retrieve a piece of information that occurs somewhere in two large sets of data. The information may be an accumulated set of totals; it may be a single event that happened once, but when is not known. If the specific piece of information is found halfway through the enquiry, the User will not want to search through the rest of the possible records when there is nothing useful left to find. If the search is accumulating totals, say, the User will not want to lose everything done so far if the system should crash, or should the User be called away. In these cases, the logical success unit would be divided into chunks of the data accessed, say 10 records at a time.

An enquiry–event pair may be a logical success unit
If an update process involves making an enquiry first, and on return of data to the screen then performing the update, the entire operation should be regarded as a success unit. If the two are made into separate success units, there is a danger that between the two the database will have been amended by another process. Thus an erroneous update is made on the strength of outdated information.

During Function Definition such enquiry–update pairs should be identified and grouped together. The role of the enquiry may be to provide the correct entry point to the database for the update, or it may prevent the User updating the wrong occurrence of the entity. When such a pre-event enquiry is found, the database should be locked from the initial enquiry through to the end of the update.

SUMMARY

Logical Database Process Design is the first of our design activities. Although Technical System Options have been selected, the design is still product independent.

There are two essential elements in the stage: producing Enquiry Processing Models and producing Update Processing Models. Both involve the use of structured diagramming techniques, and are built on SSADM products from Stage 3.

Enquiry Processing Models are built up on Enquiry Access Paths, from Logical Data Modelling combined with Input/Output Structures from Function Definition.

Update Processing Models are built up on Effect Correspondence Diagrams, from Event/Entity Modelling.

Both sets of models will be input to Stage 6, Physical Design, for the models to be transformed into code, if an application generator or 4GL is used, or into Program Specifications if a 3GL environment has been selected.

19. Physical Data Design

19.1 Aims of chapter

In this chapter you will learn the following:
- How to convert the Logical Data Model into a universal First Cut Physical Data Design.
- How to classify the target implementation DBMS.
- How to optimize the Physical Design to meet the performance objectives.
- Some common methods for implementing relationship data.

19.2 Where Physical Data Design is used in SSADM

Physical Data Design is carried out in three steps:

Step 610—Prepare for Physical Design The designer must study the target DBMS and classify its storage and retrieval characteristics. There are two forms to be completed in this task, the DBMS Data Storage Classification form, and the DBMS Performance Classification form. These should, where possible, be completed by the DBMS vendor.

Step 620—Create Physical Data Design A series of transformations is applied to the Logical Data Model to turn it into a universal first-cut data model. This model is product independent; to complete the transformation to a Physical Design we apply the rules peculiar to our selected DBMS. The particular facilities that we use in this design are selected in Step 610.

Step 640—Optimize Physical Data Design The Physical Design from Step 620 must be tested against the critical processes specified in Step 630. The aim is to achieve the performance objectives for the functions. Where this cannot be done, the design must be optimized in as many passes as are required until they are met.

Use in SSADM

The first activity in the Physical Design Module is Physical Data Design. Because it is concerned with the placement of the data model on to any one of many DBMSs, it is not possible to describe the subject in more than generic terms.

The object of Physical Data Design is the placement of the new system data on to the storage medium, in such a way that the required accesses and updates can be made within the prescribed performance levels.

There are general principles that the designer must understand to achieve an efficient placement. As the principles are more to do with database technology and traditional file organization and design techniques than with SSADM, I shall not describe them in great detail. I shall indicate the areas in which efficiency can be achieved, especially in terms of navigating data hierarchies.

Although ideally the first-cut design should achieve the system's purposes, that is very rarely the case. After producing a product-specific design, using that product's own rules, we must carry out a paper timing exercise on the critical transactions, and where we find that the model does not support the performance levels, we must tune the design in various ways until we do achieve the required levels. If that cannot be done, then we must negotiate with the User to alter the levels to a target that we can achieve.

Inputs to Physical Data Design are as follows:
● Data Catalogue
● Effect Correspondence Diagrams
● Enquiry Access Paths
● Function Definitions
● Installation Development Standards
● Required System Logical Data Model
● Requirements Catalogue
● Technical Environment Description
● Update and Enquiry Processing Models

Outputs from Physical Data Design are as follows:
● DBMS Data Storage Classification
● DBMS Performance Classification
● Function Definition (with sorting requirements)
● Physical Design Strategy
● Requirements Catalogue
● Space Estimation Form
● Timing Estimation Form
● Physical Data Design

Figure 19.1 shows the relationship of Physical Data Design to other SSADM techniques.

19.3 Procedures

Physical Data Design is carried out through a series of activities. The three step titles summarize the broad approaches of the activities:
1. Prepare for Physical Design.
2. Create the Physical Data Design.
3. Meet performance objectives.

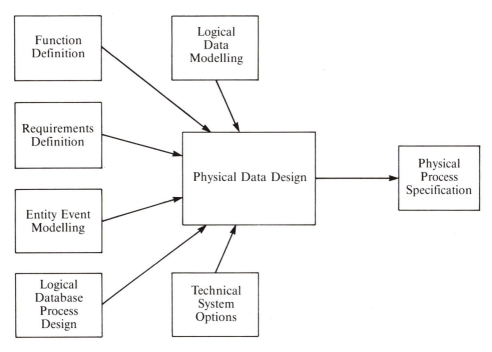

Figure 19.1

Prepare for Physical Design

Before we examine the Logical Data Model for the new system, and transform it into a Physical Design, we must first study and understand the implementation environment.

SSADM provides a means of classifying the chosen DBMS according to two basic considerations: *data storage* facilities and *performance*. There is a form for each that is to be completed. The DBMS vendor should, ideally, complete these forms, but it may well be that the designer must instead. The forms identify the particular mechanisms of storing and accessing physical groups of data that any DBMS may feature. In order to carry out the product-specific design, the designer must know and understand the particular mechanisms that the target DBMS employs. A discussion of these possible mechanisms is given later in this section.

Figures 19.2 and 19.3 give examples of the two forms using a typical relational environment. They are not connected to a specific product, such as DB2 or Ingres.

Once the classifications are complete, the designer must design a Timing Estimation form, and a Space Estimation form. These will not be completed until after the product-specific design, but as their contents depend upon the classification that has just been carried out, so they are designed at this point. Examples of these forms will be given later in this section.

DBMS Data Storage Classification

DBMS/file handler: *Relational*

Relationship representation

Table	List	Phantom
Indexes	*No*	*Serial searches*

Amalgamation of entity and relationship data

None	*Yes*
Relationship and master	*No*
Relationship and detail	*No*
Relationship and with master and detail	*No*
Relationship and relationship	*No*

Key representation in relationship (logical or physical)

Master to detail/detail to next detail	*Physical (Indexes); Logical (Phantom)*
Detail to master	*Logical (TNF data)*

Retrieval by logical key

Search	Indexing	Hashing
Yes	*Yes*	*No*

Implementation of place near logic

Clustering Indexes

Significant restrictions

None

Figure 19.2

DBMS Performance Classification

DBMS/file handler: Relational

Transaction logging overhead

For standard transactions which commit data only at termination, log overhead is minimal, as this is the only synchronous log I/O. Frequent commits increase the overhead

---

Commit/backout overhead

Commit—see above. Back-out for on line transactions, usually minimal may be substantial in batch. Processes without frequent commits, because of the need to read active log datasets (disk) and possibly archive logs (usually cartridge)

Space management overhead

Tables can be set up with free space to improve performance, particularly if data is volatile. Requirements for each table-index should be assesed and appropriate sizing adjustments made.

Dialogue context save/restore overhead

In CICS dialogue context is only saved under specific application control — no automatic features.

Standard timing factors

Disc operation:	Time:	Comment:
Read:	20 ms	Assume 3380 disk
Write:	20 ms	
Overflow overhead:		

DBMS operation	DBMS CPU time	TP monitor CPU time

An average call uses approx 15-20 000 CPU instructions, i.e. about 1 ms elapsed on an IBM 3090-180 (15 mps). DB has elaborate mechanisms to reduce syncronous I/O, especially for read only applications. Ratio of I/O requests to actual physical I/O should be about 5:1. CICS CPU time is typically minimal (per transaction).

Performance parameters for available sort packages

Figure 19.3

Once the implementation environment is throroughly understood, and the designer is familiar with the particular storage and performance features available, the strategy for the rest of Physical Design can be agreed.

These decisions may include choosing which available facilities will be used, particularly those that deal with indexing and paging, for high performance. On the other hand, the decisions may be at a higher level, and may include such decisions as whether or not to move to a product-specific design, leaving out a universal first-cut design. Decisions like this are made by experienced designers who are able to draw on that experience when carrying out the design.

All decisions made during this process should be open to revision right through the design stage, rather than be fixed and committed for ever, once made.

Create the Physical Data Design

This activity describes the first-cut design, i.e. the design has been produced following data design rules and guidelines, but has not been tested against the prescribed performance levels. As a result of this testing the design will probably need tuning until the service levels can be met. Different methods and options for tuning the design are discussed later in this section.

The objective of this activity is to create a Physical Design which matches the rules of the chosen DBMS, that can be created swiftly, and against which the required processes and enquiries can be measured.

Before we look at the algorithm for transforming the Logical Data Model into a first-cut design, we must recognize certain assumptions made by SSADM, on which the algorithm is based:

- Entities on the LDM are represented as *record types*. Each occurrence of the entity is accessed as a record.
- Records are stored on *blocks*. The block (or 'page') is the physical unit of transfer. Each record is stored on an identifiable block, so that it can be accessed. The access is through the primary key, if a direct access is required, such as a Course Run, or by being stored on the same block as a record which is accessed directly. An example of this would be storing Bookings records on the same block as the Course Run to which they relate.
- Records are *grouped* into physical groups. Records or record types that are accessed together need to be grouped together on the disk. An occurrence of a detail record, therefore, must be stored in physical proximity to its master. Such groupings should be identified in the first instance on the LDM, as it is being transformed. The data design is built around the grouping on the same block of masters and details, and details of details, etc., according to the frequency of access. The actual mechanism for so storing and retrieving the data depends on the facilities offered by the specific DBMS.
- *Primary relationships* within a physical group are supported. Primary relationships are those between master and details in the same physical grouping. The DBMS supports these relationships.

● *Secondary relationships* between physical groups are supported. Relationships between entities from different physical groups are known as secondary relationships. These are supported, but the mechanisms for this form of storage may be different from those used by primary relationships.

With these assumptions in mind, we can now begin to transform the LDM for the required system into a first-cut Physical Design. This happens in eight steps:

Step 1: Identify features of the Required System LDM required for Physical Data Design   This involves creating another model from the LDS, but omitting all information not needed for Physical Design. This model is only a working document, not the final design, nor a replacement for the LDS

This is not a mandatory step: experienced designers who are fully conversant with the DBMS may not require this model. If a software tool is available, this would be an appropriate task for it.

To achieve the transformation, apply the following guidelines:

● Turn each 'soft' box to a 'hard' box, i.e. one with square corners, to distinguish the two models.
● Remove all relationship names.
● Replace all dotted lines with solid lines.
● Each relationship which, on the Logical Data Model, is dotted at the detail end (showing that the master need not be present) should be marked on the physical model by putting an 'o' on the relationship line, just above the detail entity.
● If a master has two exclusive details, marked by an arc crossing the relationships, replace the arc with two optional 'o's on the relationships.
● Include all the design volumes from the LDS on the physical model.

This model now represents the basis for the Physical Data Design. All the business rules reflected in the Logical Data Model have been removed, as they are not a part of the physical storage plan.

Figure 19.4 shows the required LDS for SS plc. Figure 19.5 shows the LDS transformed into this first physical model.

Step 2: Identify the required entry points and distinguish those that are non-key   The entry points are derived from the set of Enquiry Access Paths and Effect Correspondence Diagrams. Identify the key access points on the model by a small circle with an arrow pointing to the entity concerned, and annotate with the key data item.

Identify the non-key access points on the model by a lozenge-shaped box linked by a crow's-foot relationship to the entity concerned, and annotate with the names of the (non-key) data items that access the entity.

Figure 19.6 shows the physical data model for SS plc with the access points marked.

Step 3: Identify the roots of physical groups   The physical groups of records are viewed as a hierarchy. The topmost entity in the hierarchy is the *root* entity. The first and most easily identified root entities are the reference entities, i.e. those that have details, but no masters.

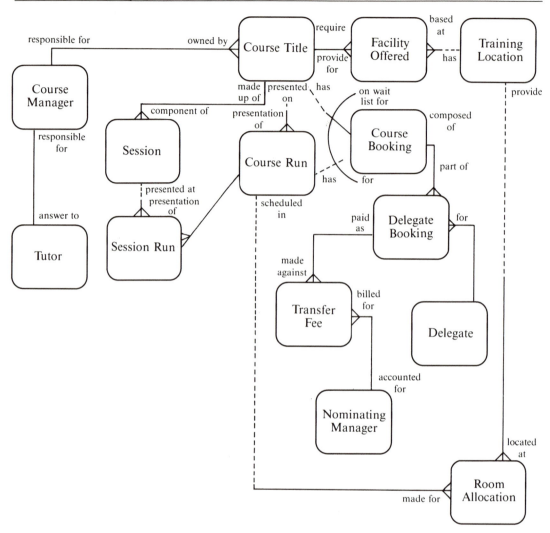

Figure 19.4

The other root entities are those entities identified as direct access points, unless that entity has a master already identified as a root entity.

All entities identified as root entities are marked with a stripe along the top.

Step 4: Identify the allowable physical groups for each non-root entity   All non-root entities should be put into a physical group, according to one of the following constraints:

● A non-root entity may only be put into a physical group if one of its mandatory masters has already been put into that group.

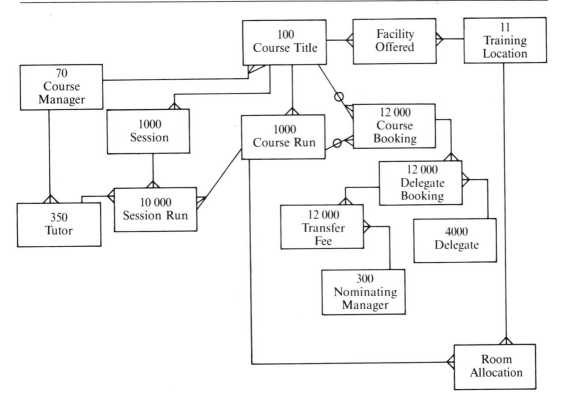

Figure 19.5

- If a non-root entity is a direct access point, and it has more than one mandatory master across different groups, put it into the group with the master whose key is part of the entity's key.

Draw lines around the groupings to show all the possible groupings.

Figure 19.7 shows the SS plc model with the physical groups marked.

Step 5: Apply the least dependent occurrence rule   Under Step 4, a non-root entity may find itself in several allowable physical groups. As it may only be put into one actual group, we apply the following measure to allocate it to just one: when an entity may be put into more than one physical hierarchy, put it into the one where it will have the least number of occurrences. This is not an infallible and inevitable rule, but a generally reliable guide.

Figure 19.8 shows the model for SS plc with the least dependent occurrence rule applied.

Step 6: Determine the block size to be used   The aim of this step is to select a block size that supports the largest of the commonly used blocks. The information can be

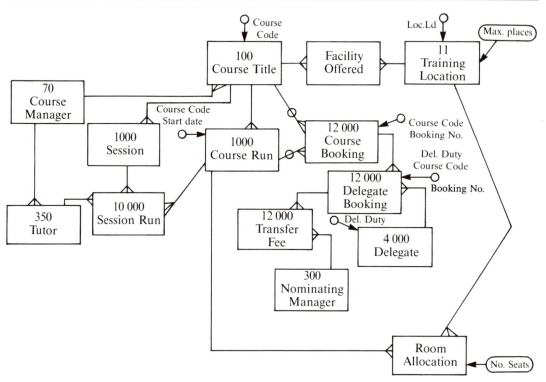

Figure 19.6

obtained from the Function Definitions. The chosen size must be supported by the target DBMS, and must not cause buffer problems when read into memory, especially when several transactions arrive close together.

When choosing a block size, therefore, take the following considerations into account:

● the block sizes that the DBMS will support,
● the size of the most commonly used physical groups, and
● the amount of space that blocks will take up when they are read into memory.

For our current example, I shall select a block size of 40K.

Step 7: Split physical groups to fit chosen block size   Once the block size is known, the physical groups must be measured against it, to be sure that each block fits. If it does not, the groups must be split in such away that they do fit.

There are two calculations to make: the size of the data in the groups, and the storage needed for relationship data:

1. To calculate the size of data storage, add the lengths of each data item for each record type and multiply by the expected number of records of that type.

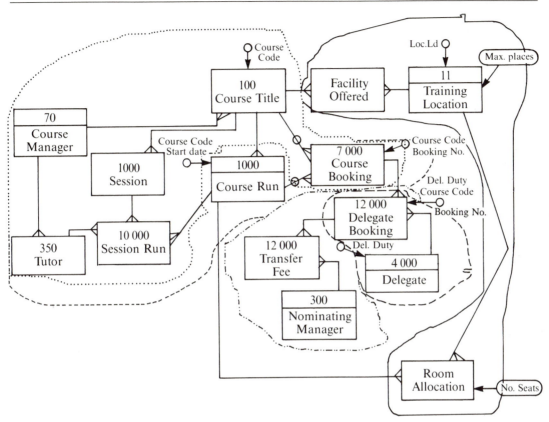

Figure 19.7

2. For calculating space for maintaining relationships and disk management, sizing standards should be available. If they are not, the rules of thumb below may help:
   (a) For each relationship, allow six bytes in the master, and six bytes in each detail.
   (b) If there is to be a secondary index, allow six bytes per entry.
   (c) Allow six bytes per record for header overheads.
   (d) Allow between 30 and 40 bytes per block for block management overheads. Assume that there will be a block packing density of about 60 per cent to reduce overflow.

Enter these estimates on a Space Estimation form. Remember that at this point these are first-cut estimates only, and will be revised considerably. Their function at the moment is to enable us to see whether our physical groups actually fit into the blocks. One of the Space Estimation forms for SS plc is to be found in Fig. 19.9.

If a group does not fit, then we must split it. We do this from the bottom of the hierarchy upwards, to find a subgroup that will fit into a block. When splitting groups in this way, re-apply the least dependent occurrence rule. If you have to do this, then

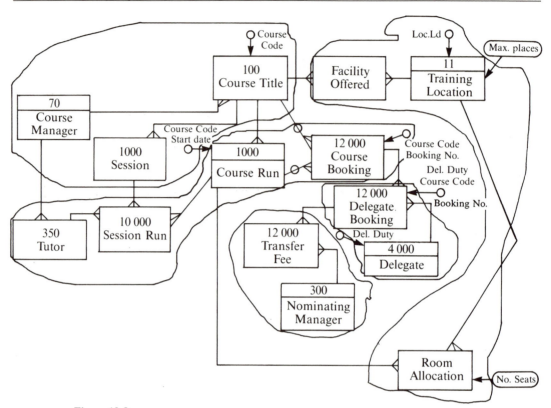

Figure 19.8

complete a fresh Space Estimation form for the new grouping. Each time you have to revisit a data group in order to fit it into the specified block size, you must update the relevant documentation. You do not need to hang on to the superseded forms; only the current valid ones are part of the deliverables.

Step 8: Apply the product-specific rules to the design   Following the DBMS Classification exercise we know about the various facilities offered by the DBMS. Now we must choose which of them to apply to our data design. Once we have chosen them we convert our universal first-cut design to a product-specific design, based upon the design rules of our target DBMS.

These rules are provided by the vendor. It may be that some products do not include such rules; in this case, the rules must be created.

In deciding upon the features to be included, and in applying the rules, there are three issues to be addressed:

1. *How are 'place-near' mechanisms implemented?* We must store the physical groups on the data model as collections of records in as close proximity as possible on the same block.

Sizing Form

| Block size: 40K | | Packing density: 60% 24K | | | Range from: 1 To: 40 | | | | |

Record	Key	Records per hier-archy	Data size	Record o/head	Primary index	Second-dary index	Space per record	Record space per block
Course Title	Course Code	1	100	36	6	18	160	160
Session	Course/Session Code	10	100	18	12	–	130	1300
Manager	Duty Code	1	50	18	6	6	80	80

Note: There is room for approx. 21 hierarchies per block; allow load space, and security requirements and we can still load 15 hierarchies.

Data total	1540
Block header	40
Block total	1580

Figure 19.9

Each DBMS has its own way of storing such records, or if they must be separated, preserving the link between them for fast access. Some DBMSs will be more flexible than others in providing for this feature.

It is up to the designer to choose the most appropriate mechanism for implementing this feature, and to what extent it will be used.

2. *What is the support for secondary relationships?* The place-near facility will affect records in primary relationships, i.e. those related to other records in the same physical group. This facility should also be made available for secondary relationships, i.e. those records that are related, but are in different physical groups. In such cases, the DBMS should provide an efficient way of moving from master to detail, or detail to master, across two physical groups. There are some DBMSs that do not cater for this facility; in such cases the designer must build one instead.

3. *What are the specific restrictions imposed by the product?* This heading can apply to almost any aspect of data storage, but in practice it usually applies to just two areas:

    (a) Restrictions on how data on relationships can be amalgamated with data on entities. This usually applies to recursive or optional relationships.

    If there are such restrictions, the designer must find a way of representing the data in spite of those relationships. Ways of doing this include creating secondary indexes, creating special records, or fields in records, to record the relationships.

    (b) Restrictions on how records can be placed near other records. If restrictions are imposed here, the designer must reduce the extent to which records are blocked together, by splitting the physical groups into smaller groups.

Meet performance objectives

After we have completed the first-cut Data Design, we prepare the Process Design in Step 630. Having defined the process, we can now, in Step 640, apply them to our product-specific model, timing them to see if the prescribed performance objectives can be achieved.

The timing and storage constraints were identified and recorded in the Requirements Catalogue during Stage 3. If the product-specific design meets those constraints straight off, then it is implemented as it is. This is rare, however, and more usually the design must be tuned, or *optimized* until it does meet the requirements.

There are two broad approaches to this: if there is a tool available locally to generate a database easily, with similar characteristics to the target environment, the timings can be tried using benchmarks. If that is the case, then SSADM guidelines need not apply.

If no such tool is available, the following SSADM considerations and activities will enable the designer to optimize the design to meet the objectives as far as is practicable.

Optimization is not a quick and easy task, and it carries overheads in the implemented design, so any such exercise should be carefully planned, with the objectives understood before it is commenced. It should be carried out by an expert in the target environment, and, where the payoffs no longer seem to justify the effort, should be stopped, and the performance objectives renegotiated.

Figure 19.10 illustrates the optimization cycle and the main activities involved. One important activity not shown in the diagram is the consultation with the User over changing the requirements, rather than continuing in the cycle until it becomes counter-productive.

OBJECTIVES

If we are to optimize the design, we must achieve one of two aims:
1. to save storage space, or
2. to save processing time.

Whichever is our objective, to achieve the one will adversely impact upon the other. In saving space we may increase processing effort, and in saving processing effort, especially to retrieve data, we may increase data storage. In either event, we are in danger of making the system rather more complex than our initial design. The danger here is that the system is less like the data model, and so less portable.

Where possible, our optimization should allow us to preserve the data model in our system, so that it will be easily maintained and more easily understood. However, there are, more often than not, times when the model cannot be maintained; this may be because of limitations in the implementation mechanisms, or because the performance of the model is genuinely too slow for our requirements.

OPTIMIZE STORAGE AND TIMING

SSADM does not give detailed instructions on how to optimize storage and timings. Instead it suggests techniques which can be applied to improve the performance of the design. To carry out the optimization properly involves a more detailed knowledge of the target DBMS, the physical environment, CPU performances, disk access speeds, access mechanisms and similar factors which are outside SSADM *per se*. The techniques mentioned in these sections represent guidelines for the optimizer to follow, using the more detailed knowledge borne of familiarity with the system.

We have already completed the Space Estimation forms to give us the storage requirements. If the configuration constraints make storage a problem, despite the low cost of backing store, one or more of the following techniques may help the situation:

* Use a more efficient hashing algorithm to distribute data more evenly across the file space.
* Compress data fields and records.
* Increase the packing density to hold more records per block.
* Increase the block size.
* Remove redundant data from master–detail relationships, and let pointers maintain the relationship.
* Reduce/remove historical data.

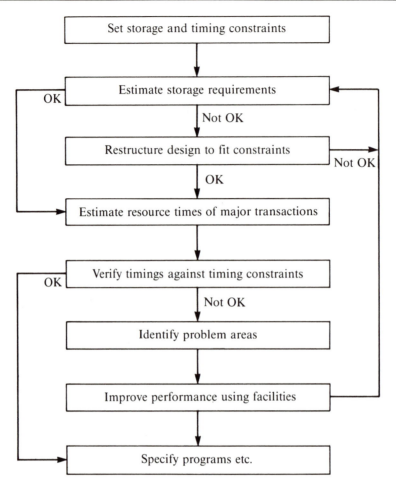

Figure 19.10

The problem with optimizing the data storage is, as I have indicated above, that it impacts upon the timing considerations. An agreed trade-off between the two must take place.

As well as investigating the storage requirements and how well our design supports them, we must also look at the resource times across the system. This is not a small task, and involves a further annotation to the Physical Design:

- Identify the major functions for timing. The bulk of the functions will not be major transactions; about 10 to 15 per cent count as major. I shall use the function Make Booking from SS plc as an example to illustrate the timing exercise.
- Time the transaction. This involves completing the Timing Estimation form we designed after classifying the system. It is completed by recording the following:

—Physical access information. In this we include the record types that need to be accessed for the function, the numbers of each record accessed and the access paths followed.

—The disk accesses required to make the physical reads and writes above. If all the details are stored by a master, there may be several physical reads through the set until the target detail is found, but only the one disk access is needed to read the whole set in with the master accessed.

—Complete the CPU time and the totals. For this we need the information we have already collected on the DBMS Performance Classification form. To complete this information, we must have the CPU figures made available to us, the translation of machine instructions through various layers of software, such as operating system, TP monitor, etc., and any algorithms or calculations for executing interrupts or scheduling.

Figure 19.11 shows the Timing Estimation form filled in for the function Make Booking.

● Identify problem areas. Having completed the timing objectives, look again at the performance objectives, and identify significant mismatches. These must be addressed and ways of improving performance devised, either by retaining the structure of the model, or by compromising the structure. In the first instance, we can use the DBMS facilities as identified, and in the second we may ignore them.

I shall not describe any of the methods in detail, but indicate particular areas that the optimizer can address.

Improve performance by retaining the structure of the design and using the DBMS facilities

There are several techniques available to improve performance, all to do with storing access information on entities, on relationships or on keys.

1. *Entity*

   The storage of entities can be optimized by any one of the following mechanisms:

   (a) Place details near the masters whose relationship is most often used. Thus when a block with the master is read in, the details it requires most are also read in.

   (b) Change the access method, so that hashed keys become indexed, or indexed keys become hashed. Obviously, questions such as 'what is the hit rate for this record set going to be for the major functions?' must be asked when deciding to do this.

   (c) Implement key-only entities as indexes, to allow a more rapid navigation.

   (d) Add direct access mechanisms to detail records, so that physical disk access does not entail reading such a large hierarchy of records before finding the target.

Timing Form

Transaction type **Insert Course Booking** Real time/batch **R**   No. of transactions: per **DAY**

Record type	Number accesses	Access type	Access path	Disk accesses		CPU time		
				Index accesses	Data accesses	DBMS CPU time	TP monitor CPU time	Application CPU time
Course Run	1	R	I	1	1	5	4	1
Course Booking	1	S		0		5	0	1
Course Booking	1	L	M		1	6.5	0	1
Delegate Booking	5	S		0		25	0	5
Delegate Booking	5	L	M	0		32.5	0	5

TOTAL

2 if update else 0   +   **3**   x37.5 **187.5** MS+   **91**   MS=   **278.5** MS

Access paths

H Hashed
C to chain detail
M to chain master
I via index
N next physical

Access types

R Read   S Store   L Link

M Modify   D Delete   U Unlink

Figure 19.11

2. *Relationship*

Add extra pointers, backwards and owner pointers especially.

3. *Key*

Read the entire pointer set into memory at the beginning of a function, and store in a look-up table. This can prevent a costly multi-pass search through sets to retrieve a relatively low number of records. This minimizes the overhead of DBMS to CPU.

There are various other ways of tuning the Physical Design that do not affect the one-to-one mapping. Common ways include:

(a) Adjust the block size, to hold more records in a single block.

(b) Increase the number of blocks held in buffer.

(c) Increase the packing density in the blocks: before going for this option, examine the likely volatility of the file first, and the impact of frequent overflow.

(d) Store the most active data on faster devices.

(e) Optimize the disk placement of files.

Where possible, it is more desirable to choose an optimization path from this section, rather than have to compromise the design. Departing from the logical model may improve performance, but it also makes maintenance of the design more difficult. Another danger in compromising the design is that ad hoc query facilities are also compromised. If the facility for such enquiries were a major requirement, then we have to forfeit a major requirement for performance in another part of the system.

Before carrying out any tuning, check with the User whether or not the performance levels achieved are acceptable. Even if the timings do not reach the specified levels, the User may find that he can work with them, and so tuning is superfluous.

Improve performance by compromising the structure of the model, or by not using DBMS facilities

If tuning is required, and the required levels cannot be reached by preserving the integrity of the model, then one or more of the following techniques may do the trick. Again, I do not propose to describe the mechanics of each in detail, but will just indicate areas where action may be considered.

1. *Redundancy*

(a) Incorporate derived data, such as results of calculations, in the masters to save processing time.

(b) Incorporate data belonging to the master record in the details.

2. *Entity representation*

(a) Merge details into masters (incorporate repeating groups into the master).

(b) Split the logical entity into two or more record types, related by key or by set.

3. *Relationship representation*

(a) Choose another way of representing the relationship other than by a set. Possible means include indexes, pointer arrays, or pointer chains.

4. *Delay updates*
   (a) Postpone all updates to records until after the transaction is completed, and they can be carried out in one pass later.
   (b) Postpone all updates which change relationships.
   (c) Postpone all updates which delete records.

In all cases the records would be marked so that logically they had been changed, but the physical activity would not take place until a subsequent occasion.

It is important to remember that tuning a design to meet the required performance levels is a costly task, and should be undertaken in a pragmatic manner. If the result is too far from the original model, or if the task of tuning is going to take an unreasonable time, then negotiations with the User should be started, with a view to altering the performance objectives.

As with the Space Estimation form, complete a fresh Timing Estimation form for each optimization path. If, for example, you choose to store master data inside detail records, there will be fewer accesses and fewer reads. The new form must reflect the difference in timing. Similarly, if you change from indexed access to a hashed access, there will be at least one fewer disk access, maybe more, and so you must complete a new form to show the timings on this path.

Complete the Physical Data Design

After tuning is complete, and designer and User are content with the results, the remaining tasks for Physical Design are completed.

Validate the impact of imposed sequencing

Examine the impact of any sorting of transactions. A sort means that the model will be hit by the transactions in a different order to that in which they occurred in the world. This may or may not present a problem. To check the validity of this new order of events, look again at the Function Definitions investigations for the function in question.

Document the sorting requirements

Where sorts are to be included, be sure that they are entered into the Function Definitions.

Identify processing optimization requirements

If the data optimizations carried out previously do not meet the performance objectives, then consider optimizing the code. This may be a matter of writing more efficient code in the same implementation language; or, if an application generator was used, perhaps some code written more efficiently in a 3GL may be needed; perhaps it may even be necessary to code a module or routine in an assembler language to achieve the required performance.

If such requirements are identified, they must be documented in the Requirements Catalogue.

Update the service-level requirements

If the performance objectives cannot be met, or if two separate functions demand conflicting objectives, the designer and User must agree on new service-level requirements. Where this happens, the Requirements Catalogue and the relevant Function Definitions must be updated accordingly.

Record optimization decisions

Any complex optimizations should be recorded for future maintenance or enhancements. SSADM does not provide a product specifically for this, but the design documentation should cater for such optimizations.

Validate the performance of the final design

Only the critical functions will be tested here. First build the Process Data Interface (PDI) Modules. These are explained in Chapter 20 (Physical Process Specification) in more detail. If the environment exists on site, they are better built using a non-procedural language. This is preferable, but not essential.

Build dummy enquiry and update Modules and submit them to a heavy series of test transactions. Monitor the performance of the database against these tests for any errors or miscalculations in the timings. If the results are satisfactory, the PDI modules can then be used in the final implementation.

SUMMARY

Physical Data Design is impossible to prescribe in strictly algorithmic terms, because the variety of implementation and development environments is so wide. If there were a standard to which each manufacturer and supplier worked, this subject could be more clearly defined. Nevertheless, SSADM defines the task as closely as possible in a generic way, always keeping the central tasks in mind: to implement the logical model as closely as is feasible, and to implement it such that performance requirements are met as closely as is feasible.

The specific requirements for data design and the hooks for SSADM are expected to be supplied by the vendor, rather than by SSADM or CCTA.

After the product-specific first-cut design, the designer enters into an optimization cycle that continues either until all performance objectives are met, or until it is counter-productive to continue. In that case, the User's agreement to change the objectives must be obtained.

20. Physical Process Specification

20.1 Aims of chapter

In this chapter you will learn:
- The purpose of Physical Process Specification.
- How to classify the Physical Processing System.
- The defining of a Physical Design Strategy.
- When to create the Function Component Implementation Map.
- When to prepare and consolidate the Process Data Interface.

20.2 Where Physical Process Specification is used in SSADM

Physical Process Design is carried out in the following four steps:

Step 610—Prepare for Physical Design In this step we must study the selected physical environment to establish how the Logical Design will be implemented. The decisions made here are all product specific.

Step 630—Create the Function Component Implementation Map (FCIM) We break each function down into its component parts. We examine each function to define its implementation plan, and to identify any fragments that can be reused.

As well as specifying for each the handling of the processes and data streams so far unspecified, we identify common processing across the Modules and in so doing remove duplication.

The end of this step is a definition of database access paths, a documented Physical Design Strategy, and specifications for reusable processing fragments.

Step 650—Complete Function Specification All specifications for functions that require procedural code are defined here. Specific function models can be drawn for any where there are divergences from the Universal Function Model. Structure clashes are resolved, and any recognized optimization requirements are specified.

Batch programs, procedural and non-procedural components will be sequenced together, along with any sort or other utilities required.

Step 660—Consolidate Process Data Interface   The fragments on the FCIM that access data are compared with the optimized Physical Data Design. Where differences are identified, extra fragments, or Modules, are specified to act as a mask for these mismatches. If a non-procedural language is available at the installation, the mask, or interface, can be written in that language at Step 660. If there is only a procedural tool available, the intervening Module will be specified here, for subsequent coding.

Use in SSADM

The purpose of Physical Process Specification is to translate the logical specification into physical programs, I/O formats and Dialogue Designs. These products will be designed for a specific environment.

Inputs to Physical Process Specification are as follows:

● Logical System Specification
● Installation Development Standards
● Physical Environment Specification
● Application Style Guide

Outputs from Physical Process Specification are as follows:

● Application Development Standards
● Function Component Implementation Map
● Function Definitions

Figure 20.1 shows the relationship between Physical Process Specification and other SSADM techniques.

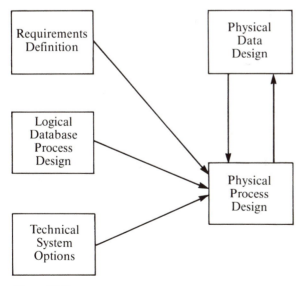

Figure 20.1

20.3 Concepts

Physical Process Design makes use of several techniques that have not been used before, and creates new products as intermediate stages in the production of Program Specifications, or Modules.

Before describing the activities of the four steps in greater detail, I shall describe the new products and concepts used in Physical Process Specification.

Fragments

Fragments is not a name for a new technique, but it is a term that will be used frequently. It describes a defined piece of processing that may exist at any level. Components of functions may themselves be broken down into fragments. A fragment must be defined in terms of purpose, of inputs and outputs and correspondences. It may be at a level with operations, or it may be to do with displaying a screen message, or defining a data group. The Function Component Implementation Map (see below) must define all processing at fragment level, as well as at full function level.

Function Component Implementation Map (FCIM)

PURPOSE OF FCIM

Much of the effort in Physical Process Specification goes into compiling the FCIM. The FCIM's purpose is to classify and specify all implementation fragments for all components. It is a network diagram that shows how the elements of the processing in Physical Design fit together, and how the logical function components map on to the physical.

FCIM has four aims:
1.   To eliminate duplicate components and fragments.
2.   To reuse common components and fragments.
3.   To specify the implementation route to be followed.
4.   To package components into success units.

COMPOSITION OF FCIM

The Function Component Implementation Map is made up of specifications of common/reusable fragments and components. As well as specifications, a comprehensive cross reference of related functions and fragments is produced.

The specifications should show the form of implementation to be used, i.e. whether it is to be implemented procedurally or non-procedurally; if non-procedurally, what languages, and what tool facilities.

Each specification should describe:
● the purpose of the procedure,
● its relationship with other procedures,
● inputs and outputs, and
● summary of operations.

Programs, or run units, also need:

- constituent programs,
- hardware requirements,
- files:
  - —media,
  - —names,
  - —sort order,
  - —volumes,
  - —volatility, and
  - —record layouts.
- transform operations,
- controls and
- checkpoint/restart requirements.

The descriptions are consistent with the specific implementation environment and data dictionary conventions.

Manual procedures, too, are defined as part of the FCIM. Features described include:

- purpose of the procedure,
- relationships with other procedures,
- resource requirements,
- management policies that impact on the procedures and
- timings.

Non-procedural specifications

There are two types of implementation path for each function component: it may be coded in a procedural language, or it may be encoded in a non-procedural language.

Where possible, non-procedural implementations should be selected. This path leads to fewer errors, and hence lower maintenance overheads. Development times will probably be lower, and greater reuse of non-procedural fragments is likely.

MEANING OF 'NON-PROCEDURAL'

In some development environments, programs and routines can be coded without the use of algorithms. Definitions of inputs and outputs is sufficient to let the tool infer the processing requirements and paths.

Database access paths, too, can be inferred by some environments, and can be generated in a non-procedural manner.

SELECTING NON-PROCEDURAL IMPLEMENTATION PATHS

Data descriptions are always coded non-procedurally, whether the description is of a file structure or a screen layout, or a report format.

If a system is simple, rather than complex, i.e. clearly defined access criteria, and has minimal processing/transformation, a non-procedural implementation is recommended. The complexity of such systems can be deduced by examining the Entity

Life Histories: if the ELHs are trivial, the system will be fairly simple, however large the data model.

If a non-procedural language is available, it should be used to specify the Process Data Interface (see below).

ALTERNATIVES TO NON-PROCEDURAL IMPLEMENTATION PATHS

If the processing is inappropriate for a non-procedural path, or the installation/environment does not support a non-procedural tool, then the process must be defined in a procedural language.

Some notation for designing the algorithm must be made, according to the Application Style Guide. To achieve consistency, the structure diagram method used to devise the LDPDs would be completely appropriate. Defining Input/Output Structures and merging them to form a process structure and allocating operations and conditions, means that after Stage 6, all that remains to be done is to code the functions.

Often, procedural environments do include opportunities for some specification of non-procedural elements. Where this is the case, the opportunity should be taken and such elements described in the component specification.

Process Data Interface

The Logical Data Model is unlikely to be mapped exactly on to the optimized Physical Data Model. The processing requirements are similarly unlikely to map exactly on to the Logical Data Model. Where such a one-to-one match does exist, the process may be implemented without problem; where there is a mismatch, we wish to preserve the logical view of the data, while recognizing that the processes will not fit on to it. An intervening Module, the Process Data Interface (PDI) must be specified.

USE OF THE PDI

The PDI acts as a mask for the Physical Data Design, so that to the process components it appears as the Logical Data Model. This eases maintenance overheads, and simplifies change as well. The Module interprets accesses specified to the database in terms of the Logical Data Model so that the correct physical data items are retrieved.

COMPOSITION OF THE PDI

The PDI is made up of Function Component Implementation Map fragment specification for DBMS accesses, and FCIM fragment specification for control utilities and non-procedural syntax.

IMPLEMENTING THE PDI

Where possible, the PDI should be implemented in a non-procedural language. This entails defining a set of 'logical views' of the database. These views are made up of:
● the required operation on the LDM,
● the requisite operation on the physical database, and
● non-procedural specification of the access paths.

If there is no facility for defining these non-procedurally, then the PDI should be specified as a set of procedural Modules that enable access to all required physical records for each operation.

20.4   Procedures

Prepare for Physical Design Processing

Step 610 is now a product-specific step. The development and implementation environments have been identified, and their facilities and constraints must be identified.

In this Step we carry out the following four activities:
1.   Study the implementation environment.
2.   Produce the Processing System Classification.
3.   Specify the Application Development Standards.
4.   Develop the Physical Design Strategy.

STUDY THE IMPLEMENTATION ENVIRONMENT

The physical environment has a number of features and a number of tools associated with it. Before we can proceed with the design or specification, we must identify the features and facilities now available to us.

The Physical Environment Description is a key input to this activity.

PRODUCE THE PROCESSING SYSTEM CLASSIFICATION

There is a form for the designer to complete that classifies the facilities offered by the physical environment. The form is illustrated in Fig. 20.2.

There are a number of issues associated with this form. To complete it successfully, the designer should be supplied with a product-specific guide from the environment vendor.

Particular issues connected with this form that the designer needs to resolve come under the broad headings of:
- flexibility,
- generation of dialogue processes,
- generation of offline processes,
- generation of physical database processes,
- automation of success units,
- automation of integrity errors,
- modularity of procedural processes,
- generation of PDI and
- distributed systems.

Flexibility

The questions to examine here concern the mixing of procedural and non-procedural specification in one function. Similarly, can online and offline processing be mixed in the same function?

Processing System Classification

Classes of tool feature   *Name of tools in physical environment*

Procedural/non-procedural   ☐ Tick if both can be provided   Online/offline   ☐ Tick if both can be mixed

Success unit

Define the alternative commit strategies that can be employed.

Error handling

Describe the flexibility and recovery contingencies.

Process components

Describe the ways in which processes can be combined and generated at run time.

Database processing

Describe DB access flexibility.

Update ☐   Enquiry ☐

I/O processing

How can different data items for different groups be related together on screen.

Dialogue processing

Describe which dialogue components can be generated.

Process Data Interface

Describe the extent to which a PDI can be generated non-procedurally.

Distributed systems

Figure 20.2

In both cases, the answer depends on the facilities provided by the environment.

Generation of dialogue processes
Questions relating to dialogue specification cover the navigation and grouping of dialogue elements, and also the mapping of physical data items/groups to displayed dialogue items:
- Can data items on screen be related from different groups?
- Can logical data groups be implemented directly as physical data groups?
- Can an error-handling dialogue component be generated?
- Can a dialogue component be generated to search for and select a specific entity?
- Can use be made of standard navigation mechanisms?
- Can these mechanisms be tailored?
- Can the dialogue navigation be specified in other ways?

Frequently, the designer will want the answers to a subset of these questions, rather than needing all facilities.

Generation of offline processes
Can the I/O processes of an offline function be generated without generating physical database processing simultaneously?

Generation of physical database processes
This issue examines the facility to generate database processes automatically, using particular features of the DBMS. If such a facility exists, it may not be suitable for the designer's purposes; alternatively, it may provide precisely what is required.
- Can physical update and enquiry processes be generated?
- Can any physical database processes that have been generated be modified or replaced by procedural code?
- Can physical database processes be generated without incurring performance overheads?

Automation of success units
Can the physical success unit be matched with the scope of a logical success unit?

Automation of integrity errors
Can the physical environment be prevented from automatically inserting integrity checks into database programs?

Modularity of procedural processes
Processes, defined logically as discrete processes, may need to be combined to run together as host/subroutine, or as shared subroutines. Questions regarding the facilities for combining such routines are:
- Can two processes be combined as host and subroutine?
- Can a process be made a subroutine of any other process?
- Can processes be combined as co-processes, and how?

- Can the data linkage between Modules be made explicit?
- Can simple copies of compiled processes be kept and combined with other processes at run time only?

Generation of PDI
Are there facilities for specifying the PDI in non-procedural language?

Distributed systems
This consideration is obviously of specialist application only. The issue is whether the distribution of a database over several locations can be ignored.

SPECIFY THE APPLICATION DEVELOPMENT STANDARDS

The local installation and the physical environment may both impose naming standards for each fragment. The standards can be supported by careful cataloguing of the fragments. This can help in identifying potential reuse, and also help in maintenance issues.

The support can consist of an indexing/classifying description:
- ID (name),
- class of fragment,
- purpose/action,
- inputs/subject/precondition and
- outputs/object/post condition.

This is a suggested description, but it is a useful way of expanding the information without extending the name length.

DEVELOP THE PHYSICAL DESIGN STRATEGY

The Physical Design Strategy (PDS) mostly involves making decisions about implementing the functions and fragments. The decision rests on whether they should be implemented procedurally or non-procedurally.

The criteria
The decisions are based on the criteria that are chosen, and the facilities identified during the Processing System Classification.

The classification form will enable the designer to answer two questions:
1. How much of the physical processing can/should be specified in a non-procedural language?
2. How far can the processes defined in the Logical Design be directly implemented as physical programs, Modules or subroutines within the Physical Processing System?

Once the decision has been made for each function/fragment, the Activity Descriptions must be customized, and the FCIM standards defined.

Apply the criteria
Once the criteria have been selected for the choice of procedural or non-procedural

implementation, they must be applied. It may be that there is no choice in the matter, through lack of facilities in the environment.

If the choice does exist, however, then a decision must be made whether to implement database accesses dynamically or in embedded code. The choice will depend on the trade-off between service-level requirements, such as: performance, flexibility and ease of maintenance.

As a rule of thumb it is best to implement update processes in a non-procedural way; the designer must be able to tailor or manipulate generated code, or to supplement it with procedural fragments.

Enquiry processes should be implemented non-procedurally. Most of them can be, with less ambiguity than update processes. The only question that affects enquiry processes is 'Is the enquiry to be a success unit itself, or is it to be bound in with an update process as a pair?'

Performance should be evaluated; if a process has been marked down for non-procedural implementation, but is a critical process which will run into performance problems, then perhaps it should be implemented procedurally, after all.

The following strategies can be implemented:

- Locate the database process code and modify it.
- Discard the database process, replacing it with procedural code.
- Implement both dialogue and database processes in procedural code.

    Whichever strategy is selected for each function, it must be well documented, including the reasons for selecting that particular path.

Generate physical dialogues

There are three possible strategies for designing physical dialogues:

1. *Input and process one input at a time*   In this strategy, the natural way of processing would be to receive one input at a time. The designer must specify a dialogue to:
   (a) present the target data group in such a way that only the one event can be input;
   (b) process the one event;
   (c) report the success or failure of the event.

2. *Input a batch of events and process each individually*   Here, the system recognizes the events input, and processes them in sequence. The dialogue will:
   (a) present the data group so that up to, say, three events can be input;
   (b) detect the data items which have been changed;
   (c) invoke up to three update process Modules, one after the other;
   (d) continue until all have been processed, even if one or two should fail;
   (e) display up to three success or error messages.

3. *Input a batch of events and process them together*   The processing of this strategy is similar to that in 2, except that if one event should fail, everything will fail, including those already processed in the batch.

    At the conclusion of this only one success message or error message will be displayed.

Define the FCIM standards

The components in the FCIM can be broken down into four distinct types, each of which should have specification standards defined for it. The four types, and their SSADM origin, are given below:

Physical processes	*Logical origin*
I/O programs	Function Definition
Dialogue control programs	Dialogue Design
Database programs	LDPD
Common Modules	Various

For each function, each of these types has a prepared Program Specification. Local Installation Style Guides to a large extent dictate the format of these specifications, but the design team must define the standards first.

Create the Function Component Implementation Map

The functions must be defined so fully now that the next task is to generate the code. If application generators are to be used in the environment, then the product of this step will be generated code. If it is to be a 3GL environment, the output will be specifications that can be passed to a programmer.

The designer is working on two levels during this step; looking for duplication and commonality across all the functions at one level, and at the other specifying the FCIM elements as fully as possible. The two levels are described more fully below.

REMOVE DUPLICATION

The areas to look at here are enquiry processes. Duplication can be identified and removed by the following:

- Removing the enquiry from an enquiry–event pair. When an enquiry process is a part of such a pair, check whether the enquiry trigger is a subset of the event data. If so, discard it.

    Check whether the enquiry process is a part of the update process. If so, again discard it.

- Creating an enquiry process by duplicating an update process. If the enquiry part of an enquiry–event pair has been created simply by duplicating part of the update process, without the operations, discard or change this. Enhancement and maintenance work is duplicated by this strategy.

    If each part is a separate success unit, the update may be rejected because of invalid states, even though the enquiry part was accepted as valid.

    If validation tests are included only in the enquiry, the data must be locked between the two to avoid intermediate change by another process; this may not be possible in the physical environment.

IDENTIFY COMMON PROCESSING

Common processes are identified early in the project. In stage 3, several common processing elements will have been identified and carried forward.

Common processes at function level or event level are not carried forward. By the end of Stage 6, the only common processes to be so implemented are at the level of database processes, I/O processes or particular algorithms that can be invoked by many processes.

Super-functions and functions
Systems can be represented as a network of functions within super-functions, database processes across functions, and I/O processes across functions. Database processes may well reference common processes, as may the I/O processes. By *super-function* I mean a set of related functions joined in a dialogue.

Offline functions can also be combined into a super-function by being batched together in an operating schedule.

Functions and logical database processes
There is a many-to-many relationship between functions and events. Each function may contain several events and each event occurrence may be contained in several functions. This leads to logical database processes being regarded as common components across functions.

Logical database processes and common processes
Common processing elements inside the logical database processes should at Stage 6 be extracted and defined as a subroutine. This includes invocations to the PDI from different physical processes.

I/O processes and common processes
Some routines that occur commonly are peculiar to I/O processing, such as altering the format of input dates, or carrying out calculations on input data, or for output reports.

Such routines that appear across the functions and I/O processes should be extracted and implemented as common modules.

Figure 20.3 illustrates this network of components, down to the common processes.

These two activities, of removing duplication and specifying common processing, encompass one level of activity during Step 630. The other level encompasses the addition of detail to all of the FCIM components for each function. There are five sets of activities to perform here:
1. Define success units.
2. Specify syntax error handling.
3. Specify controls and control errors.
4. Specify I/O formats.
5. Specify physical dialogues.

Define success units   A success unit is the portion of processing that must stand or fail as a unit. It may be just the update part of a process, or just one enquiry. For an

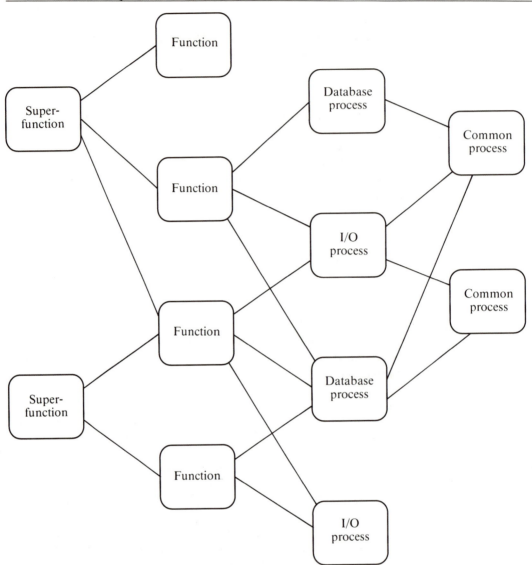

Figure 20.3

update, the success unit incorporates the presentation of data to the program and the completion of all changes of state in each entity affected. Changes to the state indicators on the ELHs indicate this completion.

The designer must define the success units based on the PDS, the Processing System Classification and the specification for each function.

Specify syntax error handling The definition of syntax error handling is not carried out until Stage 6 so that the constraints and the facilities of the physical environment can be included in the definition.

Syntax error handling is one of the opportunities for specifying reusable code: if an error is found, a code should be used to retrieve the error message from the database, rather than building the message into each and every program. This reduces the maintenance overhead considerably.

The nature of the syntax error checks is the definition of the class of data item format (alpha/numeric, etc.) and the range (e.g. valid range for numeric must be between, say, 16 and 65).

Semantic errors, or database integrity errors, are defined in the Logical Database Process Design activity in Stage 5, and so are not our concern here.

Specify controls and control errors *Control* covers two areas in this context: navigational control and control of data errors. In both cases, the classification and the Physical Design Strategy help the designer in deciding how the errors should be handled.

Specify I/O formats I/O formats are first specified during Stage 3. These are simply the I/Os for Function Definition, without consideration of error-handling. Few of these early definitions last through to Stage 6.

The I/Os relate either to dialogues and reports, or to input and output files.

In Stage 6, I/Os are designed not just for the logical I/O processing, but also for the physical error-handling and the production of reports. The logical I/O Structure will be superseded by the physical I/O Definition; the physical includes various constraints imposed by the devices that are used. Such devices include: printers, file transfer systems, terminals and backing storage devices.

The design of the I/O considers not only the data items, but also any ergonomic factors that local standards stipulate. The Applications Style Guide provides information for the designer. If external systems or devices are used, such as Electronic Data Interchange (EDI) or Bank Automated Clearing System (BACS), especial care must be taken to ensure conformance with these.

Specify physical dialogues The designer must take the logical dialogue design as a starting point, and expand it using the knowledge of the physical environment. This expansion is to incorporate such facilities as screen design and physical data groups.
1. In Dialogue Design, a physical data group is a group of data items that is displayed on the screen; it describes either the entire screen or a defined block within the screen. It may be that a physical data group matches logical data groups, one to one; this is an economic and obvious path to follow if the physical environment supports this, and if no extra update processing is entailed.

    There are four scenarios that may argue against mapping logical data groups directly onto physical data groups:

    (a) A logical data group may be too big for one physical data group.

(b) The physical environment may constrain the contents of a physical data group. In this case, obviously if a logical data group represents a greater size than can be handled, there cannot be a one-to-one mapping.

(c) The physical environment constrains the processing of a physical data group. Accesses specified on the logical model may not be supported in the Physical Design.

(d) The logical group can be used for several distinct input messages.

There are two views of data in a dialogue: the input and the display. These are not necessarily compatible. The input data relates to an event or an enquiry trigger. Each event can affect many entities, as each entity may be affected by many events. At dialogue level, a screen may display all of the data groups that may be affected by a choice of events, while the input is a subset of the data items that constitute a trigger.

2. *Physical Dialogue Design*   As described earlier, there are three possible strategies for dialogue handling: one event input and processed at a time; groups of events input and each one processed individually; groups of events input and processed together.

The Physical Design Strategy should enable the designer to choose which Dialogue Design strategy to follow. A great deal depends, as always, on the physical environment. If, for example, the data on the screen can be redefined as a number of physical data groups, the first strategy can be employed.

If, on the other hand, the environment constrains the designer to specify a data group containing more than one trigger or event, there is a decision set to follow:

(a) If the one data group is given several views—Strategy 1.

(b) If the data group can trigger several success units—Strategy 2.

(c) If the data group can trigger only one success unit—Strategy 3.

As always, the decisions must be taken only in consultation with the User.

3. *Error-handling components*   As well as enquiry and update dialogues, components to handle the three main classes of error must be defined:

(a) syntax errors in inputs,

(b) control errors in input, and

(c) database integrity errors.

Where possible, these should be generated by the environment. Where this is not possible, the logical dialogues must be extended to cater for them.

4. *Search/enquiry components*   The effort required in the logical Dialogue Design to handle searches through data sets and groups depends on the abilities of the physical environment. If the environment is sufficiently powerful and versatile, little effort is required in converting the logical to physical Dialogue Designs.

5. *Dialogue navigation*   The designer must specify the navigation paths within each dialogue, and between dialogues. Again, the physical environment may be of help. There are standard navigation paths:

(a) Move to the next/previous data item.

(b) Move forward from within one group to the next.

(c) Move back from within one group to a previous group.

(d) Move up from dialogue to menu, or to a higher level dialogue.

Logical Dialogue Design produces a set of Dialogue Control Tables. The physical environment should be able to implement these. If not, the designer must find a way; perhaps using procedural code, or perhaps by defining a physical table for the screen or for each data group.

6. *Linking dialogues—super-functions*   Menus and direct command links allow the combining of dialogues into super-functions. Menus allow a hierarchical linking. Related dialogues can be placed beside each other on a bottom-level menu, so that crossing from one to the next poses no difficulty.

It is up to the User to specify to the designer the optimum requirement for navigation and linking dialogues.

Complete Function Specification

The Function Specifications are taken from the Logical Database Process Design. Procedural and non-procedural programs must be specified. Purely non-procedural programs can actually be generated at this point. Generally, it is found that even non-procedural programs need some procedural code inserted, either through a 3GL such as COBOL or C, or through some special-purpose language.

This step must be executed by a designer. Business analyst or systems analyst skills are not the most appropriate for this technical exercise. Some of the Jackson structure clash resolutions will be the province of programmers.

PROCEDURAL SPECIFICATION

If a function is specified procedurally the processes must be specified in greater detail. The main problem is defining how the logical processes can be implemented as physical Modules, and then combined.

Options that can be considered are:

- separate logical processes in a function,
- input/output subsystems, and
- combine processes.

These are described below.

Separate logical processes in a function

The Universal Function Model in Fig. 20.4 represents a generic view of the component processes and linking data streams in a function. Functons which are specified procedurally may need this model to be adapted to reflect a complex lattice of fragments. If there is a mismatch between the logical and physical views of the data, again, a Specific Function Model helps to clarify the views.

If a Specific Function Model is compiled it should be recorded with the Function Definition, and may also be incorporated into the Program Specification.

In a procedural specification, the design is carried out using the Jackson design technique of defining input and output data structures and merging them. Chapter 18, on Logical Database Process Design, describes the possible structure clashes and how they can be resolved.

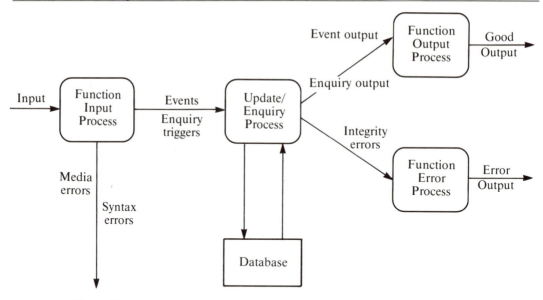

Figure 20.4

Input/output subsystems

There may be an operational or performance requirement to store inputs or outputs in a separate database, independent of the main database. This storage of the I/Os would be temporary. The inputs, processing and outputs would not take place in one success unit, and so it is not strictly a part of the function.

Where this happens, it is advisable to create a small subsystem alongside the main system to deal with the inputs and outputs. I/O Structures would define the interfaces between these systems.

Combine processes

Processes can be combined into one success unit or one complete program. They communicate by means of intermediate files. In this way, the Specific Function Model becomes a system run-chart. The use of intermediate files implies that the function specified is an offline function.

Separate processes, or *routines*, can be combined in one of two ways: co-routines, or subroutine and host. At present, few physical environments support co-routine, or parallel, processing, so the host/subroutine path is the more likely.

Generally, the dialogue process will be the host, and the database process the subroutine; this is because the database process can be executed in one uninterrupted pass as a success unit, while the dialogue process may be interrupted by I/O processing and database processing.

However the routines and processes are combined, the designer must return to the earlier specification of success units, to be sure that they do not need to be changed. If they do, then of course change them.

An important feature in designing subroutines is the data linkage. Many environments offer the facility to make the linkage explicit. If this is the case, then take advantage of it.

Consolidate Process Data Interface

If the physical environment allows, the PDI should be specified non-procedurally. Most DBMSs allow the defining of logical views of the physical database. The view consists of:

- the operation desired upon the logical data model;
- the operations necessary upon the physical database; and
- non-procedural specification of the physical access path.

If there is no such facility, then the PDI must be implemented as a set of procedural modules.

Access requirements for fragments are compared with the Physical Data Design, after optimization, and any mismatches identified.

The physical keys of master and details for each mismatch are identified, and the sequence of physical accesses determined.

The PDI elements must be fully documented in the FCIM, to show how mismatches between the two views are identified and resolved.

If low-level routines are needed for performance reasons, these must be identified and specified here. If, even then, required service levels cannot be met, then the User must agree to whatever compromise on the targets is reached, and the Requirements Catalogue should be amended to show that decision.

Assemble Physical Design

In this step, all that remains during the SSADM part of the project is to gather all the Physical Design products for validation, for checking of consistency and for final delivery. When the Physical Specification has been handed to Project Management, and has been formally signed off, the next task is to begin building the database and coding the programs. That is for another team, though. Our formal involvement in the IS project is completed.

SUMMARY

Stage 6 of SSADM can only offer generic help in specifying a Physical Design, because of significant differences between the possible environments. The CCTA subject guide on 3GL interfaces with SSADM gives more clarification on how to cope with such elements as structure clashes in the Logical Design, and how the procedurally defined processes can be translated into code, whichever language is used.

Database design is helped by the SSADM preparatory work on universal first-cut design, and on the classification of physical environments carried out once the environment is known. The suppliers of the physical environments should help in this classification work; if they do not, the designer must find out the storage and performance features directly, instead.

The optimum situation is to find a one-to-one mapping of the Logical Data Design on to the Physical Data Design. Where this does not happen, the Process Data Interface must be specified, to show the accesses required. This aids the understanding of the development, and eases the maintenance load, by requiring any change to be made to the PDI rather than the physical storage design.

PART THREE SSADM in use

21. Applications for SSADM

21.1 Aims of chapter

This chapter discusses the applications and situations that are appropriate for SSADM. The topics covered are:
- suitable applications,
- Project Procedures,
- IT support for SSADM, and
- migration from Version 3 to Version 4.

In Part 1 I described the structure of core SSADM, and its place in the project lifecycle. We saw then that SSADM is not the sole activity in the Full Study, and that it cannot take place without considerable management overhead.

In this chapter, I shall look more closely at the overheads incurred in applying SSADM, and how they interface with the method to produce a product that is as close to the User's requirements, and of as high a quality as it is reasonable to expect of any method.

21.2 Suitable cases for SSADM

Notwithstanding SSADM's flexibility, and its applicability to a wide range of projects, it is not a universal cure-all, to be applied regardless of application area. During either the Tactical Study or the Feasibility Study, the question of which approach and which method must be addressed.

There are some criteria used to examine how susceptible an application is to SSADM. They fall under the following broad headings of: data, procedures, culture, project size, management commitment and technology.

Data

As SSADM activities are centred on a stable data model, it is important that the information central to the application area can be modelled in a Logical Data Structure. If it is entirely free-form text, say, or statistics, it may not be possible to build a stable model. In that case, SSADM may not be the most appropriate approach. If a subset of the information is in one or other of these forms, but the bulk of the information is held in a more structured form, then SSADM combined with a package approach is suitable.

If the information is to be held in the form of a knowledge base, for an expert system (ES), the system should use an alternative, more appropriate method for analysis and design. ESs use a different architecture from conventional database and data processing systems. Orthodox analysis and design methods are not suitable for this particular form of application.

Procedures

SSADM expects a degree of formality in procedures in order to model transformations of data, and to identify different functions and User Roles. If investigation reveals that no such formality is present and—more significantly—is not required in the future, SSADM may be inappropriate, again. The ease with which top-level Data Flow Diagrams can be drawn indicates whether or not this is so. If clear functional decomposition cannot be carried out, then another approach, such as Checkland's Soft Systems Method, or Mumford's ETHICS would suit better. After such a study, there may be a case for SSADM to work on identified formal areas of procedure.

Culture

To succeed, SSADM must involve the Users for the whole duration of the project. The analysts and designers work in close consultation with the User Requirements from first investigation, through the whole requirements and functions analysis, to the confirmation or amendment of performance objectives.

Some organizations do not encourage such close interaction between business Users and the IS practitioners. Very often the only contacts occur during fact-finding, to approve options and to sign-off the User tests. If the target organization has such a culture, SSADM may not succeed. The culture must be one of complete cooperation between the different communities. On the one hand, the IS community recognizes the business User's ownership of the system, and so work hard to understand and meet their requirements; on the other hand, the business Users recognize their own responsibility towards achieving the IS system that they require. Thus they do not leave everything to the IS department, only to complain of the shortcomings of the system when it is finally delivered.

Project size

The cost of using SSADM techniques, with all of the Project Management and Project Procedures overheads, must be justified by the measurable benefits of choosing that path.

If the system is small (a subjective judgement, perhaps), the cost of employing the rigour and complexity of SSADM may outweigh the benefits of getting a simple system up and working fast using a simpler (or no) method.

If the system is medium to large, with a degree of complexity, the use of SSADM will pay for itself in ensuring a product delivered to a planned timescale, and to an acceptable level of quality.

A simple Cost/Benefit Analysis should be drawn up using SSADM estimating guidelines. The time to do this is either at Tactical Study level, or at Feasibility.

Management commitment

An unfortunate misconception of earlier versions of SSADM was that it was a project management method. This was never intended to be the case; the CCTA have never propounded that view of the method, neither has any other training organization. However, many managers in the industry appear to have been of the opinion that if a couple of members of a project team have attended an SSADM Course, the whole team can carry out an analysis and design project without any interaction with formal Project Management. Whenever loud condemnations have been made of the method, they frequently came from just such a manager.

As I stated in Chapter 2, SSADM occupies one part of a project lifecycle. The version of SSADM described here does not allow such misconceptions. It is not a quick and simple way of installing a system; an algorithm that can be applied by the blind and inexperienced, or the stupid. The practitioner must be trained and experienced, and have full support. Whereas this should always have been the case, I have seen examples of projects that have been resourced and planned quite wrongly.

Version 4 of SSADM makes explicit what should be common practice. The information highway and the interfaces with Project Procedures mean that an organization which uses SSADM has to commit resources to it, and plan its progress. The practitioner team is no longer the only set of actors taking part in the enterprise.

This all presumes a serious commitment to the success of a SSADM project, on the part of management. If they are not prepared to provide the full level of support, and to ensure that all required Project Procedures are in place and that the staff all have full training, then perhaps SSADM is not going to suit the culture of that organization.

Technology

Developing technologies, such as object-orientation, can be developed using SSADM, but the method must be tailored to the needs of the technology. Similarly, small systems and distributed systems are not specifically catered for in the structure and procedures of SSADM, but nonetheless can be developed with intelligent tailoring of the method. SSADM is flexible enough to cope with particular approaches and strategies of IS development. CCTA subject guides are available to describe how best to adapt SSADM to small systems, distributed systems and object-orientation.

21.3   Project Procedures

In Chapter 2 I described the two parallel streams of activity that take place during an IS project. SSADM represents one of these streams, but does not itself cover the entire lifecycle, or all of the activities. This section draws attention to some of these ancillary activities that interface with SSADM.

I shall not try to give a full description of any of these activities, or Project Procedures, in this section. The CCTA subject guides cover them in the required depth. I shall limit myself to enumerating the principal procedures, and to describing their place in the method.

Project Management

ORGANIZATION

The current structure of SSADM is designed to give optimum support to Project Management, in that it interfaces with the management using predefined products. The assumption is that the chosen Project Management method or structure will be similar to the Government standard, PRINCE. While PRINCE is a specific product, its structure and philosophy can be applied generically. Like SSADM, it is oriented towards products, rather than monitoring activities.

Broadly, a method like PRINCE employs a *Project Board*, comprising an *executive*, a *senior user* and a *senior technical representative*. Between them, these three roles represent the key interests in the project: management, the User community, and the IT community. The board's function is to control the whole project. They will be in a position to apportion resources from their own areas, as required.

The Project Board appoints a *Project Manager* who oversees the day-to-day running of the project. The Project Manager is described as the information highway in the SSADM Structure Model, acting as the interface between the practitioner team and information and control, or the Project Board.

Below the Project Manager is a separate manager for each activity, whether Module or stage. That *activity manager*'s responsibilities are to ensure that targets are met, and that products are produced to the required quality, and within budget.

The activity manager is supported by the activity team responsible for the production of the deliverables of that activity.

There must also be a *project support office*, to provide whatever support services are required, and also perhaps assist with the resourcing of the project.

Other roles are suggested within the Project Management framework, but I will not go into that issue further than to say that the above structure, at least, is necessary for an SSADM project. Local standards and culture will dictate the rest of the framework.

Figure 21.1 gives a view of such a management structure.

PLANNING AND CONTROL

The levels of management overseeing the project, either in its entirety, or at activity level, must draw up plans which will normally be in two parts: technical plans, showing the activities and dependencies; and resource plans, itemizing and costing the resources required for each activity, or phase.

Among other considerations, this is the time that the quality of each product is determined, so that it can be built into the plan. Without determining the quality criteria, monitoring the project will be less effective.

I shall not go into further detail of how planning is to be carried out, or approved: that is outside the scope of this book, describing as it does a separate activity from systems analysis and design.

Control of the plan, once the project is in progress, is another key function of Project Management. It is suggested that the control points in Table 21.1 provide the most useful controls for project assurance. The table is taken from the *SSADM*

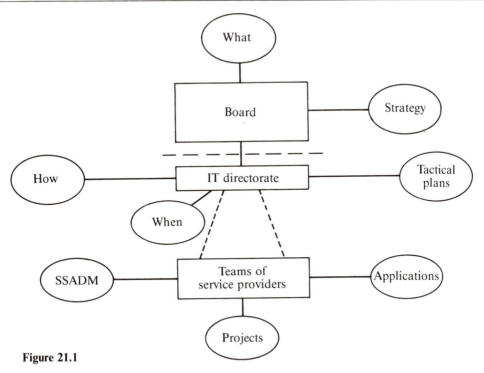

Figure 21.1

Version 4 Reference Manual. Again, I shall not describe the different procedures and issues for the controls. The CCTA subject guide on Project Management describes and discusses the issues in sufficient depth for the practitioner.

Table 2.1

Control point	Who exercises control	Triggering event
Project initiation	Project board	Authorization for project by project sponsor(s).
End module assessment	Project board	The end of a stage in the project technical plan.
Mid-module assessment (unplanned)	Project board	An explanation plan has been prepared; or the next stage needs an early start.
Project closure	Project board	All products have been delivered
Quality reviews	Review chairman	A product has been completed
Mid-module assessment (planned)	Project board	Arrival at planned point.
Checkpoint meetings	Module manager	Arrival at planned point
Preparation of highlight reports	Project manager	Arrival at planned point.

Estimating

Estimating has always been problematic in systems analysis and design, because of the variable factors such as experience of resources, availability of resources, level of staff turnover and so on.

SSADM recognizes the difficulties in accurate estimating, and by breaking down the project into chunks of modules, stages and steps, and further, by identifying products from each of these bottom-level chunks, makes it easier to estimate the efforts required to produce the individual deliverables. A subject guide to this area has also been produced, which describes the rules of thumb that can be employed. One of the most valuable aids to estimating, though, remains a manager's experience and knowledge of the project team.

Configuration Management

Configuration Management (CM) is an important part of controlling a project. The more complex the project, the more essential is good CM.

By CM, we mean a set of procedures for controlling all of the products from a project, ensuring that only current versions are produced and are in place, and that no changes can be made to any product without high-level approval and documentation. This has proved necessary after many unhappy experiences when a change has been made without authorization, and without all the knock-on effects being considered.

There are four sets of procedures involved in CM.

Configuration identification

This means identifying a product that must be signed off and controlled. Each item on the SSADM Product Breakdown Structure (Chapter 1) is a configuration item that must be subject to such control.

Such products are only placed under configuration control when they have achieved sign-off, having met the predefined quality levels.

Configuration control

Once the configuration item is accepted it is subject to both physical control, whereby the physical document is kept in a a project library, and logical control, whereby amended versions are labelled and maintained, and related sets of items are kept together as a 'release', ready to be used as input to a subsequent SSADM activity within the project.

Configuration item change

After an item has been accepted for configuration control, it may change for any one of several reasons: a new requirement by the User; an error shown up by testing, or by validation against another SSADM product, for example. The ramifications of the change must be investigated, and any other items affected changed accordingly. The change request must be sanctioned at high level, not by an ordinary team-member. As well as the technical ramifications, the project plan must also be investigated for resultant delays or changes to the critical path.

Configuration audit
Audits will be held, perhaps periodically, certainly before major decision points, to ensure that the actual products in the library tally with Project Management information on them.

Quality Control

Quality is an inherent part of SSADM, yet it does not have a fixed place in core SSADM. It is very much a feature—and function—of Project Management.

The exercise of Quality Control is aided by the establishment of a set of quality criteria for every product delivered by SSADM. There are two basic forms of Quality Control: formal and informal. Informal checks have nothing to do with management, and are not sufficient to achieve sign-off. They are valuable, however, in helping to ensure that formal quality assurance (QA) reviews are productive, and result in acceptance of the product.

Formal QA reviews follow a pattern, with actors and procedures that should be standard.

The actors are:

- *Presenter* Usually the person who has produced whatever is being reviewed. The presenter introduces the product, its scope and description, having distributed copies to all participants well beforehand.
- *Chair* Ensures that the review is conducted fairly and thoroughly. The chair may note all the comments and criticisms of the item under review, or there may be a separate scribe to do that. One main responsibility is to see that the review identifies problems and errors, but does not attempt to solve them—this can be one of the most difficult tasks, given that the composition of the review panel will be problem-solvers and experts in the functional area!
- *Reviewers* Their remit is to ensure that the items submitted are both accurate and complete, certainly up to the level laid down in the Product Decription quality criteria. They will identify errors or weaknesses, and accept ownership of them, thus ensuring that the corrections are made before the item is accepted.

Each installation will have different ways of conducting such reviews, and following up the action point, but the structure above is required for a successful review. The Product Breakdown Structure lists Quality Products as well as Technical Products. These Quality Products are all from the formal Quality Control reviews.

Risk analysis and management

The risks in question in this set of Project Procedures are to do with the security and integrity of a business's data. Just as SSADM is compatible with a Project Management style in the nature of PRINCE, so it is compatible with a risk analysis and management method called CRAMM, the CCTA's own standard. As with Project Management, the standard is not a mandatory one, but rather a generic guide.

Rather than describe the method, which is itself the subject of a CCTA guide, I shall confine myself to stating where, in a SSADM Project, Risk Analysis would be applied.

FEASIBILITY STUDY

While not a SSADM technique in itself, Risk Analysis is applied at this stage to help determine the Feasibility of a proposed project. The issues revolve around the levels of security required in a viable project.

REQUIREMENTS ANALYSIS

One of the focuses of the investigation into the current environment is the level of security achieved now, and how it is met. The findings are either carried forward to the required system as they stand, or have refinements built in to the specification for the new system.

BUSINESS SYSTEM OPTIONS

All the requirements relating to security should be incorporated into each option description, and should be submitted to the board, as a part of the supporting documentation.

TECHNICAL SYSTEM OPTIONS

The implications of required security levels are made explicit in TSOs, with descriptions of how operating software, applications software and hardware are designed or chosen to meet the requirements.

PHYSICAL DESIGN

At testing stage, the security and integrity requirements are trialled, so the testing strategy must take note of them. Similarly, the operating instructions and User manuals describe the measures to be taken.

Capacity planning

The planning in question is to incorporate a new application on to existing hardware and software, or alternatively, to investigate the impact of a new computer configuration. The configuration may be stand-alone, or it may be a part of a network—or even the whole network.

While this activity takes place outside SSADM, it requires three inputs from the core activities:

1. information on processing of data and transfers of data;
2. volumetric and frequency details, to help select a hardware configuration; and
3. information on required service levels, and non-functional requirements.

WHERE USED WITH SSADM

Step 410—Define Technical System Options   The service-level requirements and proposed volumes are tested against the suggested hardware configuration and DBMS capabilities. These are a part of the Technical Environment Description.

Step 420—Selection of Technical System Options   Only two or three TSOs are put up for selection, out of the half-dozen or so initially proposed. Once the short list is decided, the Capacity Planning techniques are applied with greater rigour, to give the selection board as much detailed information as possible, to assist with the final decision.

Stage 6—Physical Design   At the final stage, the hardware configuration and target DBMS are known, and the Logical Design is converted to map on to them. The requirements must be tested against the final design on the target configuration. Capacity Planning enables this to be done accurately, so that the design can be tuned to meet performance objectives before the final conversion and loading.

'HOOKS' INTO SSADM

The Capacity Planning exercise requires certain inputs from SSADM early in the project's life:

- Service-level requirements, taken from the Requirements Catalogue, and other products of Requirements Analysis.
- Details of all the tasks to be performed by the system, with frequencies and dependencies.
- Data storage requirements, produced by sizing and totalling all the entities in the data model, and estimating overheads for pointers and indexes. In addition, information about the clustering of groups of data is needed, and frequency of usage of data groups.
- Technical Environment Description: this provides the Capacity Planning team with basic information on proposed hardware configuration, and enables them to carry out high-level checks on the achievability of service-level requirements.

As with the other Project Procedures, the CCTA subject guide explains the techniques of Capacity Planning in detail, and describes with greater precision how and where they interface with SSADM.

Other SSADM procedures

The procedures mentioned above are a few of the most important ancillary activities. Others include:

- *Testing* at Module level, system level, user acceptance level and others.
- *Training* of practitioner staff and Users who receive the finished system.
- *Technical authoring*, using in-house standards to complete the various reports, operating instructions, manual instructions, etc.

Table 21.2 shows how they interface with core SSADM, and which way the flow travels.

Table 21.2

Procedure	Interface	Level
Project management method	I O	Modules, stages, via product/plans/ structures
Standards	I	Modules via Installation Standards
Quality control	I O	Modules stages via products/ descriptions
Capacity planning	I O	Sensibility/BSO/TSO/via option descriptions
Risk assessment	I O	Feasibility/BSO/TSO/Physical Design, via option/data/Function Descriptions
Talk-on	O	TSO/Physical Design via TSO and PD products
Physical authority	O	BSO/TSO/Physical Design, via User Requirement/option descriptions
Testing	O	TSO, via TSO/User Requirements
Training	O	Feasibility/BSO/TSO/Physical Design, via option description and design products

21.4   IT support for SSADM

Version 4 of SSADM has been designed with extensive use of IT support in mind. It would be invidious to name or recommend specific products in a book of this kind, particularly as so many new products are being developed, which will be both released and enhanced during the life of SSADM V4. I shall confine myself, therefore, to discussing briefly the generic forms of support that can be offered to the SSADM team.

The areas where IT support is most useful fall into three categories:

1.   analysis and design,
2.   Project Management, and
3.   system generation.

Analysis and design

Workbenches, CASE tools, drawing tools and data dictionaries, all provide the practitioner with valuable help. Data dictionaries, especially, should be available, whether as stand-alone products, or tied in with a CASE product. The Product Breakdown Structure provides a template for dictionary entries, while the Product Descriptions complement that with information about derivation and purpose that allows integrity checking to be built into the dictionary.

At the least, the practitioner needs access to a good drawing tool that supports the notations of SSADM, and possibly other notations that support the local Style Guide for Project Management. SSADM is very rich in graphics, and to attempt all of the diagrams on paper, and to subject to frequent revisiting and revision, would be a considerable overhead on a project.

Coupled with the drawing tool should be a good word-processor for the supporting documentation and reports that accompany the diagrams. A good CASE tool will provide both drawing capabilities with word processing and validation checks, so that a Data Flow Diagram, for example, can be validated against the rules for the technique. Ideally, cross-technique checks should be built in. Thus, the Data Store/ Entity Cross Reference would be built in, and the checker would ensure that all attributes on the LDM were created by a process on the DFD, and that every entity was created and deleted, at the least. This would also cross-refer to the ELHs, so that all three viewpoints support each other, as they do in SSADM.

Project Management

A number of Project Management tools are available on the market, and while none are specific to an analysis and design method, SSADM's hooks into management make some features especially relevant. The Product Descriptions for a SSADM project, for example, allow Capacity Planning and risk analysis and management to be automated.

The modular structure allows easier estimating to be carried out. A spreadsheet package could allow estimates to be made and tailored where necessary, and for this information to be fed into a Project Management package.

System generation

If a 3GL environment is used for development, the end-product from Stage 6 of SSADM is a specification for programs and design for a database. If a 4GL, or application generator is chosen as the development path, the end product will be a working system. SSADM Logical Design is intended to hook into a 4GL and generate the code, thus saving significant development costs. The actual interface is not defined inside SSADM, as obviously there are as many interfaces as there are products. It is intended that vendors of such products produce their own subject guides or interface specifications.

The SSADM products that help in such interfacing are the Product Descriptions, supplemented by the Product Breakdown Structure. The Entity–Relationship– Attribute Models of SSADM, published by the CCTA, provide a base for toolbuilders to design an interface for the method.

The Structural Model and the dictionary have both been designed to enable the teams using SSADM to make use of IT tools through the development cycle. The level of support actually used depends on local policies and management decisions. Basic facilities that should be used are graphics tools and word-processing. Lack of an automated data dictionary, in some form or other, will slow up the progress of the team, leading to longer integrity checks, duplication of information and increased data entry.

Ideally, a CASE tool covering most of the cycle, and made specifically for SSADM

Version 4, should be used for optimum support.

Birmingham Polytechnic is the site of the SSADM Research Centre, which tests CASE tools for conformance with SSADM standards. The Centre awards a star rating to the various products that are tested; the ratings do not reflect quality, but the range of SSADM products and facilities that are covered. Periodically, the Centre publishes findings and ratings from a tranche of products that have been tested. Any organization looking for IT support for a SSADM project should examine the findings from the Centre as the first part of its investigation.

21.5   Adoption of SSADM

There are two types of organization looking to adopt the current version of SSADM: those new to SSADM, and perhaps any analysis and design method, and those who already use SSADM Version 3, and wish to migrate to Version 4.

Those in the first category will carry out their own feasibility study on how suitable SSADM is for their purposes, and whether their culture and organization will make the best use of the method.

Those in the second category will look at two issues:

1. How can the new version be introduced, complete with the procedures and software support?
2. Can a project team currently using Version 3 migrate mid-project to Version 4?

In this section, I shall look at the second question. The first is, again, a management issue.

There are two key areas of difference between Version 3 and Version 4:

1. Architecture.
2. The move from an activity-based method to a product-based method.

There are, of course, many more differences in detail, but these are the two significant areas to address when investigating a possible migration.

The modular structure of Version 4 provides a number of entry points. Figure 21.2 shows the possible migration paths between the two versions. The end of Feasibility clearly provides a common starting point for Full Study, and the products from V3 can be input to V4 with minimal changes.

The next point where migration can occur is after the Requirements Analysis Module in V4, and after Stage 2 in V3. Not all of the products are in place to begin Logical Systems Specification, notably the data model enhanced by RDA.

The last possible migration point is between Stage 4, Data Design in V3, and Module 4, Logical System Specification in V4. If this path is selected, products from V3 Stages 2, 3 and 4 must be in place. All will be needed. The exception could be the V3 Technical Option: the first activity in Module 4 is Technical System Options. All other products from 2 and 4 must have been completed.

One place where migration will not work is from V3 Stage 5 to Module 5, Physical Design. The inputs to V4 Physical Design are very different in content and standard so that the V3 LUPOs and LEPOs would be meaningless to a V4 Physical Process Design.

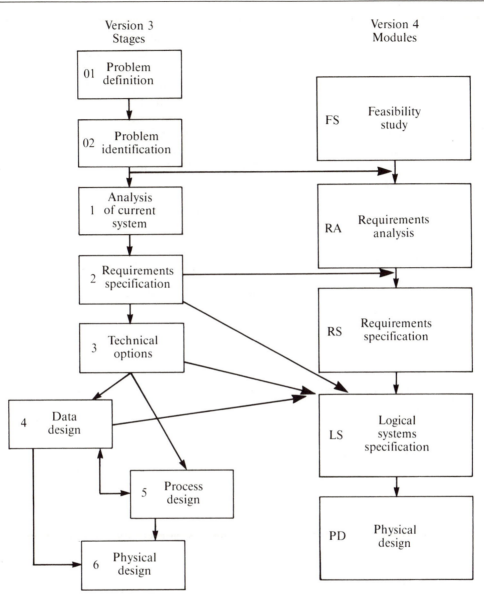

Figure 21.2

A problem with migrating, at any stage, is the difference in notation. The notation for LDM is very different across the versions, and so must be amended to meet the V4 standards. This is important with this particular technique, as the same diagram can exist in each version, but with a very different interpretation between them. Entity Life Histories will have been completed in Version 3, but not Effect Correspondence

Diagrams; ELHs will not have operations added, but may have state indicators included already. Function Definition will not have been carried out for V3, and the Dialogue Design will have been executed using old notations and philosophy. For this path to be successful, a great deal of work must be performed on converting V3 documentation to V4, and on Configuration Management.

The lesson from this is that migration is possible, but must be carefully managed. If management have made a decision that Version 4 is to be adopted, and decreed that forthwith a V3 project will be completed using V4, the Project Board and Project Managers must insist that resources are made available for carrying out an audit of products delivered from V3, and for converting the notation of one technique to another.

The Product Descriptions and Product Breakdown Structure enable the project team to make the products from the V3 part conform to V4 standards. Every product that is going to be a deliverable for the User, or which is going to be used as a subsequent input to a Version 4 activity, must be subject to Configuration Management.

Figure 21.3 shows the stages of V3 and V4 respectively, the products from one that are needed to interface with the other, and where in the project they are generated.

The other features of V4, the Project Procedures and parallel activities, should not be a barrier, or deterrent, from converting from the older version to the new: all SSADM V4 has done is to make explicit what should be standard practice in any IS development environment. If the adoption of V4 leads to a Project Management structure being installed in the organization, rather than an ad hoc arrangement, that can only be to the good.

What might cause problems would be a simultaneous adoption of V4 and CASE technology. The problem would be one of cultural adjustment, and an additional learning curve. If CASE is not already in use, this would make an excellent opportunity for introducing it. If CASE tools are installed, then the ramifications of introducing V4 need to be examined:

- Do the existing tools conform to V4 standards? Can they support the Product Descriptions?
- Can the tools be used in V3 and V4 projects until the absorption of V4 is complete?
- Can the quality criteria for products be specified and met using the tools?
- Can the tools carry out the cross-validation between the different SSADM techniques?

SUMMARY

Use of SSADM, particularly Version 4, involves a heavy commitment on the part of management. As well as the core SSADM activities, the Project Management structure must be in place, as well as the support teams providing the interfaces with Project Procedures.

Some form of computer support for the project is highly recommended, although

Stage Products

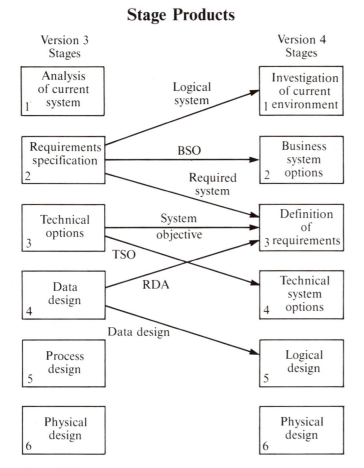

Figure 21.3

the precise form of the support, and the products chosen, is a matter for local decisions. At the very least, a good drawing package, word processor and data dictionary are required. The SSADM Research Centre at Birmingham Polytechnic carries out tests on CASE tools that claim to support SSADM, and publishes its findings periodically.

On hearing about the changes to SSADM with the advent of Version 4, clients have asked me if it takes longer to produce a system using Version 4, compared with Version 3. My answer is that it depends on how Version 3 is used; if Project Management is skimped, and the practitioner team is given little or no support, as too often happens, the Version 3 system may be delivered more quickly, but will incur heavy maintenance overheads. If the project has been conducted responsibly, and has been fully supported, the timescales for delivery will be very similar between V3 and V4. The system using Version 4 will probably be closer to the User's Requirements, because of the new emphasis on analysing and modelling User Requirements from the beginning. The specification passed to the code generators—whether human or automated—should be easier to translate into code, and so reduce both the development costs considerably, and the maintenance costs enormously.

Appendix

The SSADM user group is a body some 200 strong, with members drawn from government, industry, and academia. It has a number of special interest subgroups, such as a group for current applications, for applications using distributed systems, and many others. Any organization adopting SSADM for the first time would be well advised to join the user group for support, to draw on other organizations' experience, and to be kept abreast of developments inside the SSADM community.

The address of the chair of the user group is given below. It is correct at time of writing, although obviously circumstances may change over the years. Two other addresses of interest are given.

SSADM user group	SSADM User Group Computer Sciences Co. Ltd Computer Sciences House Brunel Way Slough Berks SL1 1XL
General SSADM enquiries	ISE Divisional Support Unit CCTA ISE Division Gildengate House Upper Green Lane Norwich NR3 1DW
SSADM Research centre	Birmingham Polytechnic Department of Computing Perry Barr Birmingham B42 2SU

Glossary

This glossary defines the terms and products that are used in SSADM. I do not classify the terms unless there is an ambiguity, so the range of expressions that follow refer to the Structural Model, SSADM techniques, products, measures or interface elements.

Acronyms are entered against the full name, not given an entry in their own right. Thus 'Business System Options' has '(BSO)' printed against it. Readers who only have the acronym must search through the initial letter entries to find it.

acceptance testing criteria   The conditions that must be satisfied for the Users to accept the system as meeting their requirements. They define the final testing process before the system is handed over to the Users.

access path   This defines the entry point into the Logical Data Structure, and the navigation from entity to entity that is needed to perform a given piece of processing.

activity   An activity in SSADM is any action that transforms a product. There should always be a description held of each activity, defining the inputs, outputs, participants and explaining what the transformation is.

Activity Network   A diagram that shows the sequence and dependencies of all activities. It is used as an aid to planning timescales and scheduling work and resources.

Analysis of Requirements   The outputs from Module 1, the Requirements Analysis Module. It comprises *Current Services Description*, *Requirements Catalogue*, *User Catalogue* and Selected *Business System Option*.

Application   An application in SSADM is that part of the business that is the object of the development work. It is defined initially in the Strategic Study.

Application Development Standards   This defines the standards by which the development of the current application is to be carried out. The standards are agreed by Project Management at the beginning of Stage 6; they are based on the *Application Style Guide* produced from *Technical System Options*.

Application Style Guide   This document is a set of standards that covers aspects of the User Interface; the standards apply to a particular development, although they are usually installation-wide. They are based on an Installation Style Guide, which should exist independently of this project, and are tailored to meet any particular requirements of this project.

attribute   An attribute is a property of an *entity*, i.e. it describes the entity in some way. For each occurrence of that entity, the attributes are set to an appropriate value, the value being drawn from a *domain*. An attribute is also known as a *data item*. An example would be:

Entity	Attribute	Value	Domain
Course	Duration	10	Number of days, between 1 and 25
	Maximum delegates	15	Number, between 5 and 20

An attribute may be 'optional', i.e. an attribute may not be given a value at the same time the entity is created.

Batch A grouping of *events* or *functions* that are performed within the same time-frame. An example would be the production of invoices. This would not be done piecemeal, one at a time, but in a batch together.

Bottom-level process A process on a *Data Flow Diagram* that cannot be decomposed to a lower level. It is marked on the diagram by an asterisk in the bottom-right corner of the box.

Business System Option (BSO) The mechanism for agreeing with the Users the scope of the functionality of the new system. BSOs are derived and selected in Stage 2 of SSADM, when the analysts prepare a number of possible scenarios for the new system. Each scenario addresses a different set of requirements from the *Requirements Catalogue*, and the *Users*, via the Project Board or similar body, will select one for development. The one chosen is known as the Selected Business System Option, and becomes the major input to Requirements Specification. The analysts help the board make the selection by preparing details of comparative costs, benefits, impacts, timescales and so on.

Capacity Planning Capacity Planning is not a core SSADM technique, but is one of the Project Procedures that supports SSADM. CCTA have produced a subject guide to Capacity Planning explaining its use, and describing methods of applying it. Its purpose is to describe the hardware and software configurations needed to meet the new system's objectives and constraints. It looks at such areas as storage, access speeds and performance prediction. It is also used as a tool to help develop the *service-level requirements*.

Cardinality *See* relationship degree.

Central Computing and Telecommunications Agency (CCTA) The Government agency responsible for coordinating and overseeing the procurement and development of computing systems throughout Government departments. Another responsibility is to recommend and implement standards in Government computing. SSADM is one of those standards, and the CCTA has been the body responsible for commissioning its production, development, enhancement and support.

　　To support the current version of SSADM, the CCTA is publishing a set of subject guides which describe activities central to an IS development, but peripheral to SSADM itself. These guides define the principal management and technical interfaces to SSADM.

Command Structure The control in a dialogue that specifies the navigational route that can be taken at the end of that dialogue. It can be terminated, or the User can move to a new one. The Command Structure permits navigation with menus or simply with commands.

Common processing Some elements of processing may be common to more than one function. A calculation, for example, may be used in more than one process; validation of inputs may be carried out in several similar inputs. Where such common processing is found, it must be documented in the *Elementary Process Descriptions*, with appropriate cross-references. They may also be carried forward in the *Function Definitions*.

Component A discrete part of a *function*, i.e. a part that can be separately identified. It is usually identified as one of the following: I/O process, database process or common process.

Composite data flow A *data flow* on a *Data Flow Diagram* which can be decomposed at a lower level into two or more separate *data flows*.

Configuration A logically related set of products requiring management and audit.

Configuration item A product that forms part of a configuration ranging from a complete system specification to an algorithm, or diagram, or piece of code. It is subject to *Configuration Management*, and as such is auditable.

Configuration Management A set of techniques to manage a configuration. The techniques ensure that the configuration items are all produced according to specified procedures for a project or local installation, and to the required quality criteria.

Context Diagram A diagram that may be produced at the initiation of a project, or at Feasibility. It defines the boundary of the system, and the major inputs and outputs that cross the system boundary. A Context Diagram may be the first product in the development of *Data Flow Diagrams*.

Control flow A management control that is exerted over any SSADM activity, whether

Module, stage or step. The control may be to start the activity, stop it or to repeat it.

Core SSADM   The five Modules of SSADM which makes up the systems analysis and design activities.

Cost/Benefit Analysis   A method for providing an objective comparison between options. The costs of developing and operating each proposed system are calculated and set off against the benefits of installing that system. It is the key supporting document for the options, both *BSO* and *TSO*, and usually carries most weight with the Project Board who make the selection.

Criteria for specification   A statement of the rules for specifying individual function components. These components may be procedural or non-procedural.

Current Environment Description   A full description, in SSADM terms, of the workings of the current system, both computer and non-computer aspects. Included is a list of identified shortcomings in its operations, and a list of requirements for the new system. If there is no existing system, the Current Environment Description will consist just of the requirements for the new system.

Current Services   A term to describe all processing that exists in the application area under investigation. The processing may be manual or computerized.

Current Services Description   The complete output from Stage 1 of SSADM. It comprises the details of the logicalization of the current system Data Flow Model, the *Logical Data Model* and the identified problems and requirements.

DataBase Management System (DBMS)   The mechanism for storing and retrieving data in a computer system. A DBMS allows data to be stored in such a way that every permitted application may use it, rather than have each application maintaining its own data files, with resultant duplication and loss of integrity.

Data Catalogue   A repository of definitions and descriptions for all *attributes* in the system, and all *domains* that give them their possible values.

data classification scheme   A means of documenting the data management facilities of the target implementation environment.

data flow   One of the elements of *Data Flow Diagrams*. The data flow shows a passage of data between two other elements, *process*, *data store* or *external entity*. At the lowest level, the data flows are simple flows, but at the higher levels they may be grouped together as composite flows for ease of drawing.

Data Flow Diagram (DFD)   A diagrammatic view of the functional view of the system under investigation. The notation comprises four elements: *process*, *data store*, *external entity* and *data flow*. Its primary purpose is to communicate with the *User* the analyst's understanding of the scope of the system, the processing undertaken and the interactions with the environment.

data item   An element of a *data store* that in some way describes or qualifies or classifies that data store. A data item equates to an *attribute* on an *entity*. Each data item on a data store must have its equivalent on a corresponding entity.

data store   One of the component elements of a *Data Flow Diagram*. It represents any place in the system where data is stored or collected. It may be an electronic medium such as a disk/tape file, or it may be a manual storage method such as a filing cabinet, or card-index file. A batch of forms in a bulldog clip waiting to be actioned also counts as a data store. On a Data Flow Diagram, it is necessary to show each data store being accessed by at least one process, and usually more, both for creating the data, and for using it.

DBMS Data Storage Classification   A means of analysing the mechanisms for storing and retrieving data on a *database management system*. This is used in Stage 6, *Physical Data Design*.

DBMS Performance Classification   A record of the factors which impact on the performance of a *DBMS*. This is used in Stage 6, *Physical Data Design*.

detail entity   One of a pair of *entity* types. Entities participate in *relationships*, in which a single occurrence of one entity type (the *master*) is associated with several occurrences of another (the *detail*). A detail entity is denoted on the *Logical Data Structure* by a 'crow's foot' attaching it to the relationship line.

determinant A term in *Relational Data Analysis* that denotes a *key data item* or group of items. The value of the other data items in the *relation* depend upon the value of the determinant. The determinant has a unique value for each occurrence of its relation.

dialogue The occasion of a User Role performing a function online. The dialogue defines the exchanges that the User has with the system via the terminal screen.

Dialogue Control Table Produced during *Dialogue Design* (Step 510), Dialogue Control Tables show the possible navigation paths in a dialogue between the *logical grouping of dialogue elements*, highlighting the sequence of aspects of the dialogue.

dialogue element A part of an input/output data flow. It may consist of just one item or it may contain several *data items*. The element is shown on the Dialogue Structure as a box.

Document Flow Diagram The initial pass at scoping a system for drawing *DFD*s. The diagram shows the flow of documents around a system, as flows from and to sources and recipients of each. Processing of these documents is not shown on this diagram.

domain A complete set of possible values from which any occurrence of an *attribute* may take its actual value.

effect The change caused to a single *entity* as a result of an *event*. The change creates an occurrence of the entity, changes the value of one or more *attributes*, including *state indicator*, or deletes an occurrence of the entity.

Effect Correspondence Diagram (ECD) A diagram to show all the *entities* affected by an *event* within the system, and how these *effects* impact upon each other. It is produced in conjunction with *Entity Life Histories* in Step 360. ECDs are used to provide the access paths for Update Processing, used in Logical Systems Design.

element Any unit or component of a *product* or *activity*.

elementary process The lowest level process on a set of *Data Flow Diagrams*.

Elementary Process Description (EPD) The supporting documentation to describe the processing on a *DFD*. Every *elementary process* on the model will have a EPD made up for it, which will describe the business activities that take place inside that *process*, and describe any decisions that need to be taken inside the process, using a tool such as a decision table, decision tree or structured English.

Enquiry Access Path The path followed through a *Logical Data Structure* in order to satisfy a business enquiry. The Enquiry Access Path is developed with the *Logical Data Model* in Step 360, in tandem with the *Effect Correspondence Diagrams*.

Enquiry Processing Model A structure diagram (Jackson diagram) that defines the processing of an *enquiry*. Operations are shown on the diagram. It is made up from the *Enquiry Access Path*.

enquiry trigger The data items that are input to define and trigger an enquiry.

entity Something about which the system needs to hold information. There should be the potential for more than one occurrence of an entity, and each occurrence of that entity should be uniquely identifiable. They are of especial interest in *Logical Data Modelling* and *Entity/Event Modelling*

Entity Description One of the supporting documents for *Logical Data Structures*. Every *entity* in the system is separately described, with all of the *attributes* and *keys* listed, as well as its purpose and use in the system.

Entity Life History (ELH) A diagrammatic technique for showing the *effect* of *events* on an *entity*. The diagram portrays all the possible events that affect any occurrence of the entity, from creation to deletion, with all the possible updates. The sequence in which these happen models the business rules governing the entity. The notation is that of the structure diagram.

entity role Denotes the situation where an *event* may affect more than one occurrence of an *entity*, but in each case the *effect* is different. Where this happens, the entity is said to have different 'roles', each role representing one of the effects. The roles must be represented on the *ELH*, as different processing is required for each role.

Entity/Event Matrix   A technique to examine all the Events that affect Data in the system, with reference to all the Entities affected by each event. It provides cross-references between LDS and ELH by ensuring that all entities are created, modified and deleted by identified events; conversely it ensures that every identified event affects at least one entity.

Entity/Event Modelling   The joint production of *Entity Life Histories* and *Effect Correspondence Diagrams.*

Estimating   One of the *Project Procedures* that supports SSADM. Producing estimates for activities and costs is a continuing activity through the project, requiring constant revision. Initial estimates are based on identified activities; these activities are weighted according to a number of factors, such as the complexity of the system, the experience of the practitioners, resource availability and so on. Estimates are made for each activity, whether Module, stage or step.

event   Something which happens in the world that causes a change in the value or status of a *data item* or *entity* in the system. It can be shown as a *data flow* responsible for updating a *data store* on a *DFD*.

External entity   A body outside the system (possibly a statutory organization such as the Inland Revenue, possibly another system in the organization, an individual or group of people). The external entity communicates with our system by means of *data flows* in or out of the system. The entity may be either source or recipient of the flows, or both. It is shown on the *DFD* as an oval with the name of the external entity inside, and a unique identifier.

Feasibility Study   The activity carried out in the first Module of an SSADM project. Feasibility is intended to define the scope of the study, and the direction it will take. During Feasibility two basic questions are addressed: can the objectives of the study be met, i.e. whether it is technically feasible and whether there is a sound business case for meeting the objectives. The study is carried out using a variety of techniques, including SSADM techniques like *Data Flow Modelling* and *Logical Data Modelling*. The report produced at the end of the Feasibility Study represents one of the User options points in SSADM, where the direction for the next part of the project is determined.

First-cut Data Design (First-cut design)   The *Logical Data Model* is subject to a transformation based on certain rules-of-thumb. These are independent of any given implementation environment, but should subsequently be adaptable to any specific product. It should be close enough to the final design for timing and sizing exercises to be carried out. There are two versions of the First-Cut Data Design: the implementation-independent design, followed by a first-cut design according to the supplied rules of the chosen environment.

fragment   A defined processing element with a specific purpose and defined inputs/outputs. It may correspond to an operation, or to a *function component*. It may support a *process*, a data group or screen message. Fragments are identified and defined during the compilation of the *FCIM*.

function   A User's view of a piece of system processing; the processing as viewed supports a business activity rather than a system or operations activity. The function is broken into smaller operational *components* for processing purposes, such as input components, output components and database read/write components.

Function Component Implementation Map (FCIM)   A decomposition of all *functions* into the *component* elements to be implemented. The components are identified in *Function Definition*, and cover the following: super-functions, *functions*, I/O processes, database processes and *common processing*.

Function Definition   A technique to identify and define the *functions* which are carried forward for Physical Design.

functional requirement   A requirement from the *User* for a particular service from the system. The service is a business request, and may be an output form, or an update process, an ad hoc query facility or a periodic report.

Impact Analysis   An examination of the effects on an organization of either a *Business System Option* or *Technical System Option*. It is one of the supporting documents produced for each option presented, and covers factors to do with organization, staffing and work practices.

information highway The mechanism by which the *product*s of each SSADM activity are submitted to *Project Management* for *quality assurance* review. All SSADM products are passed to the highway rather than on to the next activity, and must be reviewed. Any *Configuration Management* procedures in use will operate within the information highway.

Input/Output Description (I/O Description) A SSADM document to record all data about *data flow*s on the *Data Flow Diagram*. It is completed in Steps 130, 150 and 310.

Input/Output Structure (I/O Structure) The collected and documented *data items* for a given *function*. The I/O Structure comprises I/O Structure Diagram and I/O Structure Description.

key data item (or key) An *attribute* whose value uniquely identifies a specific occurrence of an *entity*. The values of all other attributes in that entity depend upon the key. In *Relational Data Analysis* the key is often referred to as a *determinant* (as its values 'determine' all other attribute values). There are several categories of key, as defined in the chapters on Relational Data Analysis and *Physical Data Design*.

Logical Data Flow Diagram The description of the current environment, after all references to physical and contingent factors have been removed. The aim is to understand the business logic underlying the current practices, rather than simply modelling the procedures in place, regardless of their validity. Typically, references to constraints caused by geographical considerations, machine-related procedures (e.g. photocopying, data prep.), office politics or office procedures are removed from the Current Physical DFD, leaving just the business logic. This takes place in Step 150.

Logical Data Model (LDM) The collected *Logical Data Structure* and supporting documentation (*Entity Description*s, Relationship Descriptions) that define the data structure of the target system. The overview LDM contains the overview *LDS* only, without the supporting documentation. That is added as the LDS is expanded. It is derived in Steps 010, 020, 110, 140, 320, 340 360 and 520.

logical data store The repository on the *Logical DFD* for all *data items* belonging to one subject. Unlike the physical data stores, where one topic, e.g. orders, could be held in many formats and many places during the life of any one occurrence.

Logical Data Store/Entity Cross Reference A cross-validation between the *Logical Data Model* and the *Logical Data Flow Diagram*. The cross-reference consists of two columns; logical data stores are listed in one column, and in the other is drawn the *entity*, or groups of entities that correspond to it, i.e. that have as *attributes* those *data items* held in the *data store*. All entities on the LDM must correspond to one and only one logical data store.

Logical Data Structure (LDS) One of the three 'views' of the system. It shows the information requirements of the organization by identifying *entities* and the *relationships* between them, in terms of the business functional activities.

Logical Design The output from Stage 5, Logical Design. It embraces the *Required System Logical Data Model*, the logical processing and *Requirements Catalogue*. It serves as the input to *Physical Design*.

logical grouping of dialogue elements A grouping together of *dialogue element*s, usually for operational reasons.

logicalization The process of converting the *Data Flow Diagram* of the current environment from the physical description to one reflecting the business logic only. The process involves rationalizing the *process*es and *data store*s.

Logical Process Model The collection of all processing details incorporated in *Logical Design*.

master entity The *entity* at the 'one' end of a *relationship*. Of the two entities in a relationship, one will be associated with one and only one occurrence of the other, while that other may be associated with more than one occurrence of the first. The entity that may occur many times is known as the *detail entity* and the one that occurs once only is the master entity.

Menu Structure A diagram that shows a hierarchy of menus within an online system, with the possible navigation paths.

Module   The significant division of SSADM activities, for management purposes. A SSADM project is composed of five Modules, each of which is made up of one or more *stage*s. Each Module produces a defined set of *product*s, and performs a defined set of *activities*. The Module is deemed to be complete when the products are completed to predefined *quality criteria*.

Non-functional Requirement   A requirement of the new system that describes the service levels or security levels that it must meet. It meets a performance need, rather than a fundamental business need.

normalization   A synonym for *Relational Data Analysis*, a way of transforming unstructured data into minimal logical groupings, such that each *attribute* is grouped only with its sole *determinant* and other attributes that share that determinant. The aim of normalization is to ensure that every entity is in the condition known as *Third Normal Form*.

offline function   A *function* that is performed by the computer without terminal interaction with the *User*. The trigger for the function may be input online, or else in *batch* mode, but all the database processing will be carried out subsequently, without reference to the User.

online function   A *function* that is performed by the computer while the *User* is at the terminal; the system and the User will communicate with input and output messages while the function is being executed, the contents of the input messages being influenced by the preceding output messages.

operation   Discrete pieces of processing which, together, constitute *effect*s. These are identified initially during *Entity/Event Modelling*, and are entered on the *Entity Life Histories*. Subsequently they are expanded on the *Update Processing Models*.

optimization   A tuning of a database model to enable it to meet the performance requirements of the specification. Optimization is a broad term to cover several kinds of activity, that range from changing the *Logical Data Model* to altering the physical placement of *data items* on the disks, to amending the way that the program logic works.

Outline Current Environment Description   A product of *Feasibility Study*, it describes the current services in the environment, and the problems associated with their provision. The description is at a high level, rather than a detailed level, and includes overview *DFD*s and *LDS* of the current system.

Outline Required Environment Description   Another product of *Feasibility Study*, this describes in broad terms the scope and requirements of the new system. As well as text description it may include overview *Data Flow Diagram*s and *Logical Data Structure*.

parallel structure   An element of *Entity Life History* notation. It denotes possible updates to the *entity* occurrence that can happen at any time during its existence, without a predetermined sequence, and without affecting the business cycle impacts on the entity. Reference data such as 'Address' on a customer record will be affected by parallel lives.

Performance Requirements   A set of *Non-functional Requirements* that define the performance of the new system. The areas covered include *response time*s for online systems and turn-round time for offline systems, as well as recovery specifications.

Physical Application Specification   The output from *Physical Design Module*. It is a collation of all the documentation that makes up the technical design.

Physical Data Design   The database definition which is to be implemented as a result of the optimizing of the *Physical Data Model*.

Physical Data Model   The *Logical Data Model* after it has been transformed in Stage 6 by applying first a non-specific set of rules for transforming it, and then applying the rules of the target DBMS to that model. Until it has been optimized in one of several possible ways, it is unlikely to meet the *Performance Requirements*, and so is not yet a *Physical Data Design*.

physical environment   The selected environment on which the new system is implemented. The term embraces DBMS, with its classifications of storage and performance. Also included are details of hardware products, software development environment (i.e. 3GL, 4GL and so on) and other software such as TPMS, Operating System, etc.).

Physical Function Specification   A full description of the processing requirements of the new system.

physical key   A device to locate the target record on the physical disk. It is also known as a 'pointer'.

problem definition statement   The definition of the User Requirements for the new system. It is produced during the Feasibility Study, and incorporates charts and diagrams.

procedural model   A structure diagram with operations and conditions appended. It is used as the formal definition of the procedural processes.

process   An *element* of a *Data Flow Diagram*. It records any transformation or manipulation carried out on a *data flow* or on *data item*s.

Process Data Interface (PDI)   Produced during Physical Design, it shows how the logical processes that access the LDM are mapped on to the physical processes that access the physical database. If there is a one-to-one correspondence between logical and physical accesses, then there is no need for a PDI; if the two do not match exactly, the PDI shows how the mismatches can be resolved in terms of physical access. If a non-procedural language, such as SQL is used at the site, the PDI can be implemented in just one or two lines. The aim of the PDI is to allow the designer to implement the logical update and enquiry processes as programs, independently of the physical database structure.

Process/Entity Matrix   A tool to help group bottom-level *process*es during *logicalization*. It is also used to perform a cross-validation between *LDS* and *DFD*, by ensuring that every *entity* is affected by at least one creation and one deletion *event*.

Process Specification   The output from Step 360. It comprises the following assembled products: *User Role/Function Matrix*, *Function Definition*s, *Required System LDM*, *Entity Life Histories* and *Effect Correspondence Diagrams*. The development of this product may identify errors in the component parts (e.g. there may be some *event*s on the ELHs never reflected in the Effect Correspondence Diagrams). When this is the case, a revisiting of the relevant products and techniques is called for.

Processing System Classification   Classifies the details of the processing environment that will be used for implementation. It frequently defines the development environment, too.

product   Any item of documentation, software or hardware that is produced during a SSADM project. The product itself may be made up of component products, such as the *Process Specification*. The *Product Breakdown Structure* classifies all products under three broad headings: Management Products, Technical Products and Quality Products.

Product Breakdown Structure   A hierarchic structure that itemizes all SSADM *product*s, and categorizes them under three broad headings. Management Products encompass such documents as Project Initiation Document, Application Development Standards, Highlight Reports, Project Plans, Module and Stage Plans and Post-implementation Review. Technical Products include Application Products, Operations Products, Education Products and Release Packages. The products from systems analysis and design activities come under the heading of Application Products. Quality Products refer to all aspects of the continuing *quality assurance* process. The products include invitations to *Quality Control* reviews, records of each review, and records of follow-up action required.

Product Description   A specification for every *product* listed on the *Product Breakdown Structure*. *Project Management* are responsible for completing these as part of the input to every *activity*. Information listed in the descriptions include derivation, composition, quality criteria and external dependencies.

program   A set of codes that fulfils the requirements of one or more *functions*.

project   A set of *activities* that together present the following features: a defined and unique set of technical products that meet the business needs; a corresponding set of activities to produce those products; nominated resources; plans; objectives and deadlines.

Project management   A set of techniques and organizational structure to plan, monitor and control the conduct of a project. It is concerned with the monitoring of the project resources and the observation of any constraints.

Project Procedures   The tools of *Project Management* that support the activities of *core SSADM*. They include such activities as *Risk Analysis*, *Capacity Planning*, *Estimating* and *Configuration Management*. They are handled by specialist teams outside the SSADM practitioner team, and interface via Project Management.

prototyping   A method of providing the *User* with a simulation of how the system will work. The simulation is modelled and demonstrated on a specialist tool that will imitate the features of the implementation environment. It is used in SSADM to confirm or clarify the User Requirements.

Prototype Pathway   A document showing the sequence and combinations of *dialogue elements* and screens demonstrated in a *prototyping* session with the *User*.

quality   A measure applied to all SSADM *products* to show that they are fit for the required purpose. Every product is classified according to standard criteria, one set of which is the *quality criteria*; if the product meets these criteria, it may then be accepted as part of the project documentation.

quality assurance   A set of mechanisms that collectively guarantee that all *products* from a project meet the required standards.

Quality Control   The process of ensuring that the required *quality criteria* are built into the products. This can be done by various forms of inspection or review during the course of production and before acceptance of a product.

quality criteria   Prescribed characteristics of a given product; these characteristics are specified before the production begins, and are passed to the practitioner team via the information highway.

quit and resume   A feature of *Entity Life Histories*. It is a device to show that particular *events* can cause a change in the prescribed sequence, in certain circumstances. It is a notation that describes how processing is stopped at one point, and resumed elsewhere.

random event   An *event* which may occur at any time during the life of an *entity*, rather than at a predetermined stage in its life.

relation   A group of *data items* under its logical *determinant*. It is represented as a table of entries, each row entry being an occurrence of that *relation*. In *Relational Data Analysis* each relation is regarded as an *entity* for the purposes of validating the *Logical Data Model*.

Relational Data Analysis (RDA)   A technique for analysing data to its logical groupings. A set of 'raw', i.e. unstructured data, is reduced through a set of transformations (*normalization*) to basic data groups in which each item is grouped with its sole and full *determinant*. The effect of performing RDA, as a step in data design, is to reduce data redundancy, and gain flexibility. During the normalization process, it is important that information is never lost, i.e. that all the original connections between data items are maintained. The output from RDA is a set of *relations* that are in the state known as *Third Normal Form*.

relationship   An association between two *entities*. A relationship that is identified on a *Logical Data Model* must hold true for all occurrences of those entities.

relationship degree (cardinality)   The class of number of each *entity* taking part in a *relationship*. At one end of the relationship there may be one and only one occurrence of an entity type. This entity type is the *master* in the relationship. At the other end there may be many occurrences of that entity type. That is called the *detail*. Relationship degree classifies the ends of the relationship as being one-and-only-one, and many.

The degrees, or Cardinality, are described in the following terms:
- $1 : 1$ one to one
- $1 : m$ one to many
- $m : n$ many to many

Required System Data Flow Model   The version of the Data Flow Model that describes the processing in the new system.

Required System Logical Data Model   The version of the *Logical Data Model* that shows the structure of the data for the new system, and the Information Requirements.

requirement   A feature of the new system that has been requested by the *User*. The requirement may be functional (i.e. to perform the application environment's primary activities) or non-functional (i.e. service requirements such as specified response-times for online activities, or recovery/security requirements).

Requirements Catalogue   The central document that lists all *requirements* from the new system. It is started at the beginning of the project, and is maintained through all stages until Physical Design is finished.

Requirements Definition   A procedure running through the early parts of a SSADM project. There is discussion over whether this counts as a technique or a procedure. It focuses on the provisions of the required system which will meet the *User*'s business needs. In the later parts of the project the identified *requirements* will be resolved by techniques such as *Function Definition*.

Resource Flow Diagram   A component part of a *Data Flow Diagram*. Its purpose is to show the physical movement of goods in a system, rather than the movement of information. Its function is to help with the initial scoping of the system, and to help identify flows of data in the early stages of the analysis.

response time   The time taken for the system to respond to a *User* input, in an online activity.

Risk Analysis   A set of procedures to identify risks to aspects of the system. These aspects may be 'hardware', security, software, physical installation, or any area where a threat is perceived. As well as identifying the risks, the practitioners must assess them, and prepare contingency measures to counter the risk. The aim is to reduce the risk to an acceptable level, but at an acceptable cost.

Service-level Requirement   A *requirement* for the new system, but one at an operational level, rather than a business function level. It describes the expected, acceptable level of service, in terms of *response time*, and recovery times.

Space Estimation   A special form to assess the storage requirements of the data design for the target implementation environment.

Specification Prototyping   A technique employed in SSADM to validate selected *User* Requirements by prototyping certain dialogues and procedures. Specification Prototyping is not used as an incremental way of building a system, only as a validation exercise with the users.

structure diagram   A diagrammatic notation for structures or procedures, shown in terms of sequence, selection and iteration. These three basic constructs are sufficient to model the structures richly. They are based on Jackson diagrams, used in Jackson Structured Programming (JSP), and Jackson Structured Design (JSD).

stage   A unit of *activity* within a SSADM project, with defined input and *products*. It is a subunit within a Module.

state indicator (SI)   A feature in *Entity Life History* analysis. SI consists of a non-functional *attribute* (single digit numeric) in an *entity*. Every *event* that affects that entity causes a change in the value of the SI. A birth event sets it initially to value = 1, and a death event sets it back to value = null. The SI can be used as a validation or referential check: if an SI is not set to a valid value for the event, the process will be aborted and an integrity error flagged. In this way, all the updates to the entity can only happen in the correct sequence.

step   A subunit of *activity* within a *stage*. Like a stage it has its prescribed inputs and *products*.

Structural Model   The model of SSADM that describes the Activity Descriptions and architectural description, to define the activities carried out in the SSADM part of a project.

success unit   A set of processing which must succeed or fail as a whole for each input. If there is a failure and the processing aborts, the system is restored to its state before the success unit began. The success unit may be a complete update, a complete enquiry, an enquiry/update pair, or a predefined number of database accesses in an enquiry, depending upon the *function*'s requirements.

system   In SSADM terms, the totality of the technical output from the project. The system exists outside the boundary of the project, and exists—as a result of the project—long after the project is complete. The system will have existed before the project as well, if it is an enhancement project.

Technical Environment Description (TED)   The description of the technical environment, after the selection of the *Technical System Option*. The description is a major input to Physical Design.

Technical System Option (TSO)   TSOs describe the technical implementation of the *Requirements Specification*. It is prepared and selected in Stage 4: Technical System Options. The *stage* has two *steps*: 410, Define Technical System Options and 420, Select Technical System Option. The option/selection procedure is similar to *BSO*. In addition, the team preparing the TSO needs to receive information from the *information highway* on Risk Assessment and *Capacity Planning*. The documentation supporting TSO is considerably more detailed than that for BSO, and following the adoption of one solution, tenders can be invited on the strength of those documents and figures.

Third Normal Form   The output from *Relational Data Analysis*. A *relation* that is in Third Normal Form is expected to be in the most logical grouping, not to be subject to update anomalies, and to give the least data redundancy. The way to achieve Third Normal Form is to examine all Second Normal Form relations for dependencies between non-key *data item*s.

Timing Estimation Form   A form designed during Stage 6, Physical Design. It is designed after the target DBMS or file system has been classified, and is intended to show how long a defined transaction will take to be performed, according to the database design and the facilities of the target DBMS.

transient data store   A store of data that is held temporarily before being processed, and is then deleted. Such data is not likely to be included on a *Logical Data Structure*. It is mostly used for administrative convenience, and frequently vanishes from the *Data Flow Diagram* at *logicalization*.

Universal Function Model   A graphical representation of a SSADM *function*, showing how the *product*s from earlier *stage*s are built into the *components* of the function for processing.

Update Processing Model   A structure diagram that defines the processing sequence of an update transaction. It is a synthesis of *Entity Life Histories*, with the associated *Effect Correspondence Diagram*s, and lists the operations on the database that are carried out during the processing.

User   The person/job that will make use of the IT system to perform a business task.

User Catalogue   A document identifying the *Users* of the online components of the system. It incorporates job titles and job descriptions.

User Role   A collection of *Users* who share particular tasks or functions.

User Role/Function Matrix   Used in the identification of dialogues, the User Role/Function Matrix is a grid that marries up *User Role*s with *online function*s. Where there is a match between the two, a dialogue is required.

Solutions to exercises

Chapter 9   Old Krate's

This solution describes a Level 1 current system DFD. The main points to note are:
1. The duplication of external entities.
2. Labelling of data flows, especially double-headed data flows.
3. The bottom-level process (Process 2).
4. The data flow into D3 Provisional Bookings from process 2: although no data is being written to the data store, it is being amended by a deletion. Therefore, the flow must still go into the store.

Process 2 has an Elementary Process Description. The other processes are decomposed to identify the contributory tasks and data stores accessed. For instance, process 3, Pay Agent's Commission must look up agents' contracts, tax tables, etc. in calculating amount to pay, as well as simply totting up the business they have put in Old Krate's direction.

Because process 3 is carried out monthly, the trigger is shown as a data flow from the data store, rather than a flow across the boundary.

A Data Catalogue will be opened for all the data items in the data stores.

Old Krate's

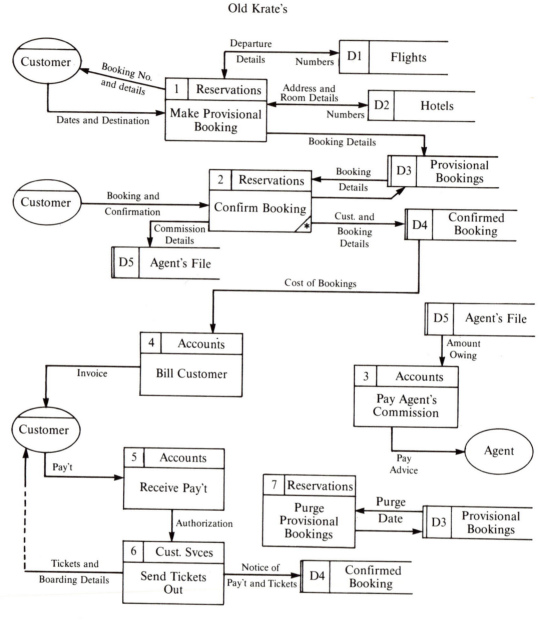

Figure S.1

Chapter 10

1. (a)

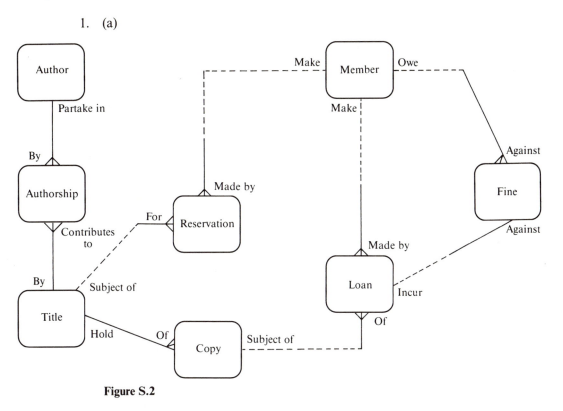

Figure S.2

The entity Loan acts as a link entity between Member and Copy. While link entities are not required in the current description to resolve many-to-many relationships, in this case it is required as a real entity in its own right, even though the environment description does not mention Loan at all.

Blacklist is mentioned, but does not qualify as a candidate entity as there is only one occurrence. There are many entries, so Blacklist Entry might be regarded as an entity. Further investigation reveals that it is an attribute of Member, a Blacklist Indicator, so it is not shown on the diagram. It will be shown on the model though, in the Entity Description of Member, and with its own Attribute Description.

Fine is in a one-to-one relationship with Loan. This is acceptable for a current system model, but in a Required System LDM it would have to be resolved. The resolution could be either to include it as an attribute of Loan, or to treat that relationship as a one-to-many. Relational Analysis in Step 340 may help the decision.

(b)

(i)

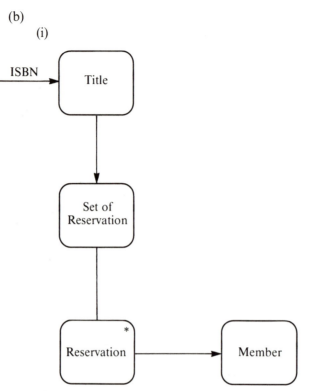

Figure S.3

This is a simple EAP, involving a short navigation across three entities. As we are looking for all reservations for that title, we need to build the extra structure box Set of Reservations. The correspondence is between Title and Set at the top level, and between Reservation and Member at the bottom level.

(ii)

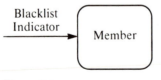

Figure S.4

This is an even simpler EAP, involving one entity and one search criterion. An EAP is no less good for being simple!

(iii)

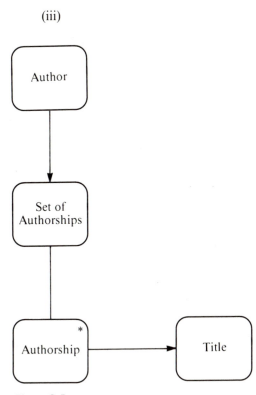

Figure S.5

This is similar to (i), involving a simple vertical, then horizontal, navigation and just three entities on the model.

2. (a)

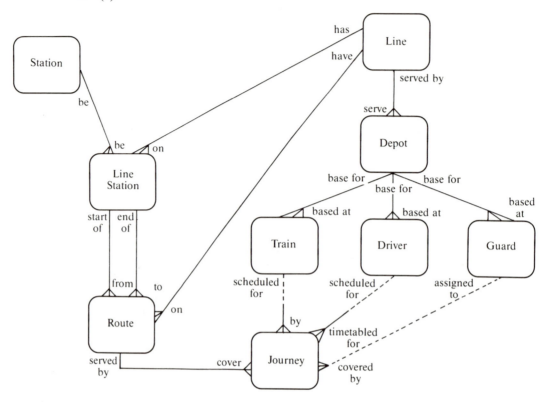

Figure S.6

This is a more complex diagram. The main point to notice is that the relationship between Line Station and Route is a double one—the start of a Route and the end of a Route. In such a case, each relationship must be drawn on the model, and each relationship will have its own pair of Relationship Description Forms.

There is one relationship which is optional from detail up to master: Journey to Guard. That is shown by the broken line connecting entity to entity. Most of the other relationships are mandatory.

(b)

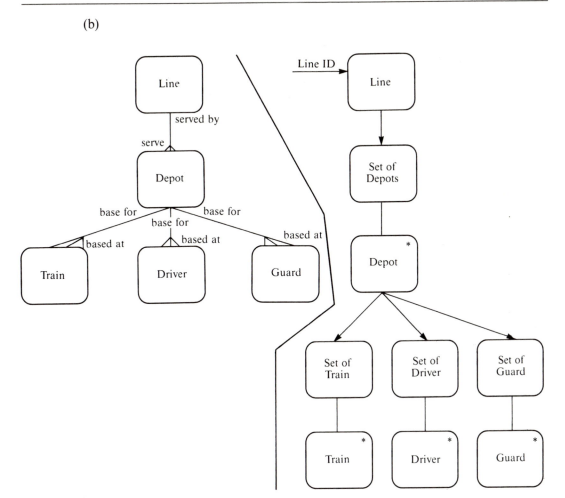

Figure S.7

This is another straightforward EAP. The navigation is all downwards, i.e. master to detail, and preserves the hierarchic shape of that portion of the LDS.

Chapter 12

This exercise extends to Fourth Normal Form. There are no instances of a 5NF projection. As it happens, all of the relations broken out in the 4NF projection are absorbed into others at optimization. This is not an inevitable occurrence; in many cases the 4NF projections must remain, as they contain information not otherwise obtainable. Opposite is the full set of normalized relations, showing the progress from UNF.

System: Hospital rotas

UNF	1NF	2NF	3NF	BCNF	4NF	5NF	OPTIMIZED
Ward Name Date Ward Type No. Beds Nurse No. Nurse Name Nurse Exp. Pat. No. Pat. Name Bed No. Date/Adm. Ded	Ward Name Date Ward Type No. Beds Sister Ward Name Date Pat. No. Pat. Name Bed No. Date Admin. Ded Ward name Date Nurse No. Nurse Name Nurse Exp.	Ward Name Date Sister Ward Name Ward Type No. Beds Ward Name Date Nurse No. Nurse No. Nurse Name Nurse Exp. Ward Name Date Pat. No. Bed No. Pat. No. Pat. Name Date Admin. Ded	Ward Name Date * Sister Ward Name Ward Type No. Beds Ward Name Date Nurse No. Nurse No. * Ward Name Nurse No. Date Ward Name Nurse No. Nurse Name Nurse Exp. Pat. No. Date * Ward Name Bed No. Pat. No. Pat. Name Date Admin. Ded	Sister Date * Ward Name	Ward Name Date Nurse Name Date * 4NF relations projected from	No 5NF relations Nurse No. Date Ward Name	**Sister** Ward Name Date * Sister **Ward** Ward Name Ward Type No. Beds **Ward Staff** Nurse No. Date Ward Name **Nurse** Nurse No. Nurse Name Nurse Exp. **Patient Ward** Pat. No. Date * Ward Name Bed No. **Patient** Pat. No. Pat. Name Date/Adm. Ded Sex * Consultant No.

'Sister' turns out to have 'Nurse No.' as its key.

Relationships must be noted if Nurse is in role of Sister.

Figure S.8 (part i)

System: Hospital rotas

UNF	1NF	2NF	3NF	BCNF	4NF	5NF	OPTIMIZED
Consultant No. Date Specialism Pat. No. Pat. Name Ward Name Bed No. Sex	Consultant No. Date Specialism Consultant No. Date Pat. No. Pat. Name Ward Bed No. Sex	Consultant No. Date Consultant No. Specialism Consultant No. Date Pat. No. Date Pat. No. Ward Bed No. Pat. No. Pat. Name Sex	Consultant No. Date Consultant No. Specialism Consultant No. Date Consultant No. Date Pat. No. Date Pat. No. Ward Bed No. Pat. No. Pat. Name Sex	Pat. No. Consultant No. Ward Name Date Pat. No. Bed No.	No 4NF relations	No 5NF relations	**Bed Allocation** (Ward Name) (Bed No.) Date * Pat. No. **Consultant Duty** Consultant No. Date **Consultation** Consultant No. Date Pat No. **Consultant** Consultant No. Specialism **Sister** Sister Date * Ward Name

Figure S.8 (part ii)

The model built up from the optimized relations is shown below.

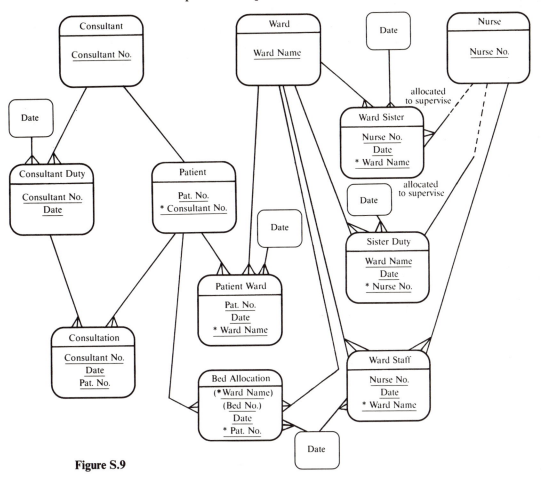

Figure S.9

Chapter 13

1. The ELH for Loan is straightforward: there are no parallel lives involved, and no necessary use of quits and resumes. As the main life is composed of selections in the iteration, rather than a sequence of events, there is no need to quit out of other activity when the book is returned: that event has its own natural place near the end of the cycle.

    Although up to three renewals and three reminders are permitted, it is not necessary to specify a separate event for each of the possible three. The attributes on the record are single fields with a specified range of values. When the Update Process Model is drawn the condition 'If renewal indicator not > 3' and so on, can be specified.

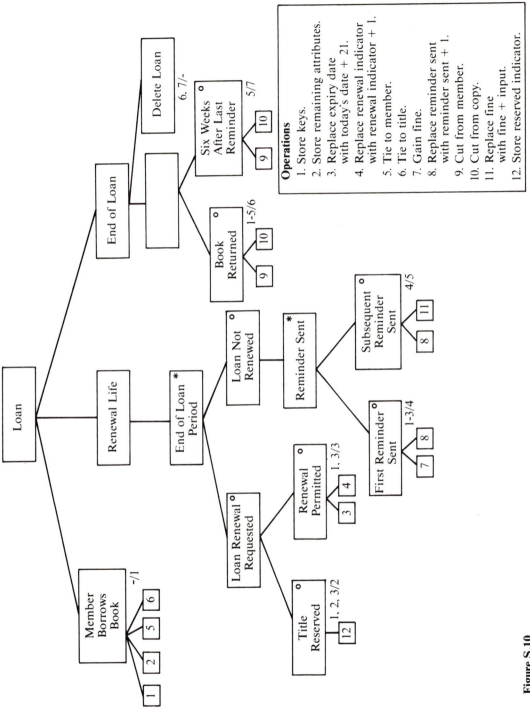

Operations

1. Store keys.
2. Store remaining attributes.
3. Replace expiry date with today's date + 21.
4. Replace renewal indicator with renewal indicator + 1.
5. Tie to member.
6. Tie to title.
7. Gain fine.
8. Replace reminder sent with reminder sent + 1.
9. Cut from member.
10. Cut from copy.
11. Replace fine with fine + input.
12. Store reserved indicator.

Figure S.10

349

2. The Effect Correspondence Diagram is, again, a simple one: the only entities affected are Loan, Member and Copy.

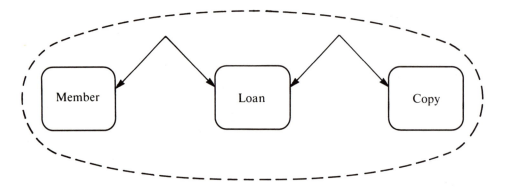

Bibliography and Selected Reading

The source for all information on SSADM Version 4 is: CCTA (1990) *SSADM Version 4 Reference Manual*, NCC Blackwell, Manchester.

Other sources of information regarding the techniques and philosophies used in SSADM are listed below.

Checkland, P. (1981) *Systems Thinking, Systems Practice*, John Wiley.

Codd, E. F. (1970) 'A relational model of data for large shared data banks', *CACM*, **13**, 6.

Codd, E. F. (1974) 'Recent investigations into relational database systems', *Proceedings IFIP Congress*.

Date, C. J. (1986) *An Introduction to Database Systems: Volume 1*, 4th edn, Addison-Wesley, Reading, Mass.

de Marco, T. (1980) *Structured Analysis and System Specification*, Prentice Hall International.

Fagin, R. (1983) *Acyclic Database Schemes (of Various Degrees): A Painless Introduction*, IBM Research Report RJ3800.

Fagin, R. *et al.* (1982) 'A simplified universal relation assumption and its properties', *ACM TODs*, 7, 3.

Gane, C. and Sarson, T. (1978) *Structured Systems Analysis: Tools and Techniques*, Prentice Hall Inc.

Jackson, M. A. (1975) *Principles of Program Design*, Academic Press, London.

Jackson, M. A. (1983) *System Development*, Prentice Hall International, London.

Loomis, M. (1983) *Data Management and File Processing*, Prentice Hall International.

Martin, J. and C. Finkelstein (1981) *Information Engineering*, Savant Research Studies.

Mumford, E. (1983) *Participative Systems Design*, Manchester Business School.

Yourdon, E. (1989) *Modern Structured Analysis*, Prentice-Hall International.

Yourdon, E. and L. L. Constantine (1979) *Structured Design*, Prentice Hall, Inc.

Index